Constance Reid

Courant

in Göttingen and New York
The Story of an Improbable Mathematician

Springer-Verlag   New York • Heidelberg • Berlin

Library of Congress Cataloging in Publication Data

Reid, Constance.
    Courant in Göttingen and New York.
    The Story of an Improbable Mathematician
    Includes index.
    1. Courant, Richard, 1888–1972. I. Title.
QA29.C68R44   510'.92'4[B]   76–17062

With 36 photographs.

© 1976, by Springer-Verlag New York Inc.
Printed in the United States of America.

ISBN 0–387–90194–9   Springer-Verlag New York
ISBN 3–540–90194–9   Springer-Verlag Berlin Heidelberg

Korf erfindet eine Uhr,
die mit zwei Paar Zeigern kreist,
und damit nach vorn nicht nur,
sondern auch nach rückwärts weist.

Korf a kind of clock invents
where two pairs of hands go round:
one the current hour presents,
one is always backward bound.

From *Christian Morgenstern's Galgenlieder*,
translated by Max Knight,
University of California Press, Berkeley, 1963.

GÖTTINGEN, a town situated 67 miles south of Hannover and, since the Second World War, a part of the state of Lower Saxony in the Federal Republic of Germany. It first appeared in written records of 953, received municipal rights in the early 13th century, and played an important role in the Hanseatic League during the following century. Still surrounded by most of its ancient ramparts, it has a medieval town hall, several Gothic churches, and a number of half-timbered houses from the 15th and 16th centuries. It is best known for its university, founded in 1734 by the Elector Georg Augustus of Hannover (George II of England) and famous since the time of Gauss for its scientific faculties. The presence of such mathematicians as David Hilbert, Hermann Minkowski, and Felix Klein at the beginning of the 20th century attracted students from all over the world. After the First World War, Göttingen quickly regained its position as a great international scientific center under the leadership of the mathematician Richard Courant and the physicists Max Born and James Franck. This preeminence came to an end in 1933 when Franck resigned and Courant, Born, and others were removed from their positions by the Nazis. The majority of the Göttingen mathematicians and physicists emigrated ultimately to the United States. . . .

ALPHA AND OMEGA

I N THE SUMMER OF 1970 I received a letter from K. O. Friedrichs, a professor at the Courant Institute of Mathematical Sciences in New York City. I had sent him a complimentary copy of *Hilbert*. He wrote to thank me for it, and then added in a postscript:

"My wife and I, and many of our friends, have often urged Courant to write his reminiscences, but he has never responded to our suggestions. I wonder whether you would feel like approaching him on this matter. I know that he reacted very positively to your work on Hilbert. I can also assure you that you would find a great deal of support among all his friends."

I already knew the story of how Richard Courant, one of Hilbert's best known students, had been removed by the Nazis from his position as director of the internationally famous mathematics institute in Göttingen and how he had emigrated to the United States and built up another great mathematics institute. When I was again in New York, I went to talk to Friedrichs and his wife in his office at the Courant Institute on the corner of Fourth and Mercer streets, a few blocks from Washington Square.

Friedrichs had been Courant's student and assistant in Göttingen and his colleague at New York University for many years. It was apparent that he was fascinated by the contradictions of Courant's character as a mathematician. "Think of it—a mathematician who hates logic, who abhors abstractions, who is suspicious of 'truth,' if it is just bare truth!" He emphasized to me, however, that he did not have in mind another biography. "Courant is no Hilbert, and he would be the first to tell you that." But Courant was on the eve of his eighty-third birthday. For most of his years he had been in close personal contact with nearly all the important mathematicians and many of the physicists of the twentieth century. He had also been a part of the great migration of the 1930's which had shifted the center of the scientific world from Europe to the United States. It would be regrettable if Courant should die without communicating to a wider audience than his own colleagues and students his recollections of the events and the personalities of his time.

Friedrichs did not minimize the difficulties of getting Courant to commit himself to such a project. Courant, he told me, often seemed incapable of saying yes to anything.

Mrs. Friedrichs was more optimistic.

"Since we first came to this country in 1937, whenever Courant was by himself at home we have usually asked him to dinner. And in all that time I don't think he has ever accepted. Although he has always come."

1

Friedrichs had not yet spoken to Courant about the project he had suggested to me, but he had arranged with Mrs. Nina Courant that I would come out to New Rochelle the following Sunday and have lunch with the family and some of Courant's friends.

A sign on the front door of the big old-fashioned wooden house said simply "Courant." Inside, the rooms bulged with heavy German furniture—many cabinets, desks and chests, all overflowing with books and journals and papers and pictures and music. There was a grand piano in the living room and several stringed instruments stood near it.

A portrait of Hilbert hung over the fireplace in the dining room. It had been painted from photographs after his death in 1943 and was no great example of the art of portraiture; yet, better than any photograph of him I had ever seen, it managed to convey his pure and intense intelligence. At the table, counting Courant, there were four members of the National Academy of Sciences, all of them German-born, all of them originally from Göttingen.

Courant himself had aged since I had first met him in 1965, when I was working on *Hilbert*. His small figure was even stouter through the middle. Increasing deafness had affected his sense of balance and he moved warily. Because of his deafness he also found it hard to follow the general conversation at lunch. With a dozen people around the table, he was quiet; but he listened intently, head bent forward. When someone spoke to him, his eyes became bright with interest and sheer enjoyment in what he was being told. "Ja, ja," he would say softly, nodding his head and encouraging the speaker on.

He listened to Friedrichs's proposal without comment or suggestion, but he did not reject it.

Friedrichs was encouraged. There at the table he began to plan how I would return to New Rochelle after Courant's birthday on January 8 and spend a week as a guest in the house. Courant could talk to me then about what he remembered and what he considered important, and we would simply see how the project developed.

Courant never actually agreed to Friedrichs's plan; but, saying goodbye to me, he said something almost inaudible about wishing me success with my project.

I was back in New Rochelle in the middle of January 1971, and Courant and I began to talk. He was fascinated by my small cassette recorder and commented how wonderful it would have been to have had such a thing in Göttingen during the First World War when Felix Klein was delivering his famous lectures on the history of mathematics to a small group gathered around his study table.

Looking back at the transcribed pages of our first conversation, I see that Courant always referred to our project as "this situation I find myself in."

"I have been urged several times to write down my reminiscences and collect material about my life and to talk into a machine like yours and have somebody transcribe what I said," he told me. "But I have always been much too lazy or undecided. Then Friedrichs came and said he had talked to you and maybe this would be much easier. I have wanted all the time to write something about the academic life and the sciences in Göttingen, which I was a witness of and which was a very great time. But this is not quite the right thing anymore. You have done it very well in your Hilbert book. But still, of course, there are things I would like to talk about. I have followed up some things consciously. I have no ambition to have any sort of biography of myself. But still it seems reasonable for us to talk about a few things. We do not have to start at alpha and end at omega."*

During the year 1971, Courant and I talked many times about the things he had done, the people he had known, his opinions and feelings about mathematics. I think he rather enjoyed our conversations, and he willingly gave me access to papers and letters in the house in New Rochelle and at the institute. But it very soon became clear that I had come too late for the project which Friedrichs had had in mind. Courant had neither the vigor nor the desire to go back over his life meaningfully. Maybe he never could have, for—as Nina Courant said to me—"Richard was always forward-looking."

Felix Klein, whose place he had taken in Göttingen, had spent his last years compiling his collected works—"a masterpiece," Courant told me—each important work or group of smaller works prefaced by a biographical note which set it in the framework of Klein's life and times. But, much as he admired what Klein had done, Courant could not bring himself to do something similar. He took comparatively little satisfaction in his past achievements. He was concerned about the future of mathematics and of the institute which he had created, and he was frustrated and unhappy because he could no longer help. Not only did he lack the physical and mental energy, but mathematics had passed him by.

* Omissions and editing necessitated by the nature of taped conversation have not been indicated in the text.

All during the year that I talked to him he was deeply depressed.

On November 19, 1971, he suffered what at first appeared to be only a slight stroke. He was taken to the hospital in New Rochelle; and on January 27, 1972, a few weeks after his eighty-fourth birthday, never having left the hospital, he died.

The book which Friedrichs had proposed that I write—Courant's reminiscences—would never be written; yet it seemed to me that I had material for another book about Courant.

In our conversations I had been constantly struck by how important his youthful experience in Göttingen had been for him. The days as Hilbert's student and assistant had molded his view of mathematics and of science. The example of Felix Klein had guided him throughout his career as an organizer and as an administrator. Everything he had done professionally, even the writing of his books, had been influenced by these two men. But he had also put his own mark on the scientific tradition which they exemplified for him—first in Germany and then later in the United States.

I was warned that while Courant's reminiscences would have been welcomed by the scientific community, a book about him would be looked upon by many as a posthumous example of the self-aggrandizement, both of himself and of his group, for which he was so often criticized—the ultimate chutzpah. I was told that nobody who came in contact with Courant was ever able to remain neutral about him. Many people, including Friedrichs and his wife, were deeply attached to him; but there were a number of other people who disliked or even detested him. All of his life he had attracted controversy. Almost every piece of mathematical work he had ever done had brought him into a questionable position in relation to some other mathematician. Some of his best work had been in collaboration with his students and had evoked murmured references to *le droit du seigneur*. His books—the importance of which is never denied—were often reputed to have been written largely by his assistants. His way of doing things made many people uncomfortable. He was sometimes referred to behind his back as "Dirty Dick" or "Tricky Dicky."

Even some of his most loyal admirers at the Courant Institute expressed doubts about the project I proposed. "A book about Courant, sure, but it should be written fifty years from now when everybody here is dead." They were afraid that a book at this time might have of necessity to be a piece of personal and institutional puffery. And then, they pointed out, there was the impossibility of the subject, Courant himself. Almost every statement one could make about him could be matched by an opposite statement equally true. I was told by Lipman Bers, a professor at Columbia who was formerly at the Courant Institute, that he had once proved "by contradiction" the theorem: *Courant does not exist.*

In spite of the many objections, I continued to feel that an account of Courant's activities on behalf of what he conceived as "the Göttingen tradition" was the natural and necessary sequel to my book about Hilbert— the rest of the story. Although it was not the book which Friedrichs had originally proposed to me, he was willing to help; and one of the great pleasures of my life has been working with him.

The book was a long time in the writing. I think that Courant would have appreciated the fact that now, five years later, when I have finished the story of his relationship with Göttingen, I find that, if I have not written his biography, I have written the story of his life.

ONE

COURANT never volunteered much about his life before he came to Göttingen.

He told me that he had been born in Lublinitz, a small town in Upper Silesia, now Polish but then German, and that his father had been a small businessman from a large Jewish family "with a not very intense intellectual life." He also said that he had supported himself since he was fourteen. He never went back in any detail beyond that year. Children of his colleagues remembered that they had been warned by their parents: "Uncle Courant does not like to talk about his childhood."

I already knew something about the Courant family history, having read the autobiographical memoir of Edith Stein, the Catholic philosopher-martyr, who was Courant's cousin on his father's side. According to her, their grandfather, Salomon Courant, was originally a soap and candle maker. Coming to Lublinitz to hawk his wares, he saw and fell in love with twelve-year-old Adelheid Burchard, the daughter of the owner of a cotton factory, a very religious man who had earlier been a cantor. Salomon returned to Lublinitz each year until Adelheid was eighteen, then married her and settled in the town. It was a county seat, the marketing and shopping center for the largely Polish peasantry of the surrounding country-side. Working together, Salomon and Adelheid developed a prosperous "food and feed" business, produced fifteen sons and daughters who survived infancy, and—after the birth of Courant's father, the sixth child and second son—built a large house where, according to Edith Stein, Salomon "practiced unlimited hospitality." Adelheid died before Courant was born, but Salomon lived on until the boy was eight—"a small lively man always full of good humor and good ideas who could tell jokes by the dozens."

Little would be known of Courant's immediate family and the early years of his life, even by his wife and children, if it were not for another, unpublished memoir written by his father, Siegmund Courant, to justify to his sons the details of his unsuccessful business career. According to this memoir, Siegmund worked in the family business from the time he was fourteen years old. He felt that he worked hard and long but that, even when he was thirty-four and about to marry Martha Freund, he had no authority in the business. Everything was decided by his older brother, Jakob.

Martha was the daughter of a successful businessman in nearby Oels. At the age of nineteen she had come to work in the store of "S. Courant" as a clerk (a slender red-haired girl, Siegmund told his sons) ; but at the end of the first month, for some unstated reason, Jakob had sent her home to her father and brothers "with twenty marks, a railway ticket and a chocolate bar." Later it was suggested to Siegmund that she might be a good wife for him. He traveled to Oels and looked over the Freund business; her father came to Lublinitz and looked over the Courant business. Terms were negotiated by Jakob. Martha was to bring a dowry of 15,000 marks in cash, and her father was to take over the mortgage on the big Courant house.

After their marriage in Oels on February 17, 1887, Siegmund and Martha went to live in Lublinitz in the house of old Salomon, where several of Siegmund's brothers and sisters were already maintaining separate apartments. For some reason his sisters were very critical of Martha. He complained that she even had to cook some of the things she liked in secret.

On January 8, 1888, Martha gave birth to Richard. He was the first grandson. Everybody in the family was delighted to have a boy to carry on the Courant name.* There was a gala week-long celebration. But Martha continued to be scorned by the Courant sisters. A year later she produced another son, who was named Fritz. There began to be friction in the family about her father's rights. Siegmund, still frustrated by his inferior position in relation to Jakob, decided to leave Lublinitz. He sold his share in "S. Courant" and purchased a business in Glatz, a town in the foothills to the west of Lublinitz. That same year, when Richard was three, a third son, Ernst, was born.

In Glatz, Siegmund and Martha had a large apartment. Relatives from both sides of the family came for pleasant visits. Although somewhat undercapitalized, the new business went well. But Siegmund was burdened

---

* The derivation of the name is not known. Edith Stein suggests that the family may have lived originally near the French border. Friedrichs thinks it more likely that the name derives from the term *Thaler courant*—the current value of the *Thaler*—an expression frequently used in money exchanges.

by interest on loans from the family and assessments made, he complained, without consulting him. When his father died, he inherited only his gold signet ring with the *S* on it. He arranged to sell the Glatz business and contracted to buy a more modest one in another town. Jakob insisted that he abandon the purchase and come instead to Breslau, where other brothers and sisters were by then in business. Since Jakob promised to pay any judgment rendered against him, Siegmund broke his contract and went.

Richard was nine when his family made the move to the Silesian capital, now also Polish. In Glatz he had already completed one quarter of the gymnasium. In Breslau he was enrolled in the König–Wilhelms Gymnasium, a humanistic gymnasium which prepared its students for university careers. His first report card ranked him twenty-first in a class of twenty-seven and bore the comment that his achievement in arithmetic was barely satisfactory, "or a little less than satisfactory."

In Breslau the Siegmund Courants found themselves in severe economic straits. They had a reputation for being "wasters," and they could not live on what Siegmund earned from a job he had obtained with an insurance company. He established a connection with a wholesale paper goods firm. The man whose business he had been going to buy sued him for breach of contract.

Richard shortly overcame what was apparently poor preparation in Glatz. By the time he was eleven, his ranking in his class oscillated between first and second, his grade in arithmetic was "very good." In the early spring of his fourteenth year—disturbed by his parents' mounting financial problems—he answered a newspaper advertisement for a tutor "qualified in all subjects." Although he got the job, he lost it very soon. The boy he had been hired to teach was lazy and slow. Richard was impatient, then angry, then abusive. The boy's mother finally slapped him and threw him out of the house. It was a cold, wet day. Courant never forgot—this he told me—how he found himself lying in the mud with a tiny gold piece in his hand, the first money he had ever earned.

In spite of this initial pedagogical failure, he soon got other tutoring jobs, becoming quickly more patient and more effective. He tried to follow the example of one of his gymnasium teachers, a man named Erich

Maschke, who was an exponent of the "Socratic method" of leading students to discovery on their own. Courant told me that he was no relation to Heinrich Maschke, who was for many years a professor of mathematics at the University of Chicago and who was also from Breslau. Maschke, the gymnasium teacher, was not a scientist and not even too well informed in mathematics. Courant recalled him as not knowing more than the elements of the differential calculus and as being hazy on the integral calculus. But he loved science and had the ability to inspire scientific excitement in his pupils. Over the years these included, in addition to Courant and his classmate Wolfgang Sternberg, such scientists as Max Born, Ernst Hellinger, Heinz Hopf, and Otto Stern among others.

While Richard was beginning his teaching career, it was becoming clear that Uncle Jakob was in serious financial trouble. The details are confusing. According to Siegmund's account, Jakob had become involved in various "shady deals" and had signed notes in the names of Siegmund and Martha without their knowledge. In the chapter of her memoirs entitled "The Sorrows of the Family," Edith Stein wrote that Jakob's own activities were blameless and that his difficulties were due to the fact that he had become involved in the business ventures of Siegmund and another brother. I was told by Helen Pick, a Courant cousin who now lives in the United States, that her parents later explained to her that there had simply been a tragic misunderstanding. At any rate, in September of Richard's fourteenth year, Jakob shot and killed himself. There was a general breakdown in the Breslau business of the family. Everyone, without exception, placed the blame on Siegmund and Martha and refused to countersign the note which would save their business.

"At a family council," Siegmund reported to his sons, "we were read out of the family."

A month later he declared himself a bankrupt.

While Siegmund was having these financial difficulties, Richard was earning a good amount of money for a boy of his age. One of his early pupils had been Wolfgang Sternberg's sister Grete, who was a student at a recently established private gymnasium for girls. The mathematics instruction there was exceptionally bad; and before long a number of Grete's classmates followed her example and came for tutoring to little Richard Courant, who was two years younger than most of them and a year behind them in school. In 1972, in Berkeley, I had an opportunity to talk to Magda Frankfurter Frei, who had been one of these girls. She remembered Richard Courant as "a very good teacher, who didn't tell you right away what you had to do but led you to understand it by figuring it out for yourself."

"What kind of a young man was he?" his cousin Helen Pick mused, repeating my question. "Very nice. Very. A little bit peculiar, you might say. You'd ask him something and he'd always circumvent it somehow, making jokes and so on. I remember my brother and me sitting together and listening to him tell stories. He always spoke in a very soft voice. So you had to listen hard or you would miss the joke. It was our greatest pleasure. I think he invented the stories because he saw what pleasure they gave us."

Among the girls whom Richard tutored, there was one who was especially good in mathematics. This was Nelly Neumann, the daughter of Justizrat Neumann, a leading member of the Jewish community in Breslau. Nelly was encouraged in her scholarly interest and ambition by her father. He was "an exceptionally kind and noble man," according to Edith Stein, and, although a Jew, looked like "a Germanic aristocrat." Since his wife's death, when Nelly was two, he had had a very close relationship with his daughter. "The happiness of their life together was disturbed only by his mother-in-law, who had remained in the house after his wife's death and continuously tortured him and his child with her bad humor."

Nelly's classmates never visited her because of her termagant grandmother. Richard, coming to give her mathematics lessons in the evening, was probably the only young person to enter the Neumann house; and in time a special friendship developed between him and Nelly.

After the bankruptcy Siegmund's account of his business activities is difficult to follow. He constantly assures his sons that "everything was in order." But the next two years were miserable. He was charged with distributing obscene postcards and fined fifty marks. Although he appealed the conviction, it was later upheld. Martha gave birth to twin sons, who survived only a few hours. In a period of six months the family lived in four different places.

During Richard's sixteenth year, his parents decided to leave Breslau and go to Berlin. He was earning enough money that he could elect not to accompany them. He rented a room and—except for a few months when his brother Fritz stayed with him—lived alone, attending classes during the day and going from house to house in the late afternoon and evening to give mathematics lessons.

Siegmund proudly emphasized in his memoir that throughout all the trouble with the Courant brothers and sisters Richard remained staunchly on his parents' side. After the rest of his family went to Berlin, the only relative he saw was Aunt Auguste Stein. A widow with eleven children, she had successfully taken over the management of her dead husband's lumberyard. Although she had sided with the others against Siegmund, she had acted as the intermediary between him and the rest of the family.

"To know of a shadow on her father's name and the conflict between her brothers was very hard on her," wrote Edith Stein, her youngest child. "And if she did not see her brother for many years after that, she showed his children a ready interest and sympathy and was happy that all of them became hardworking people who made up with industry what their parents had neglected in their education."

Aunt Auguste's favorite among the Courant boys was Ernst, the youngest; but she also admired the precocious Richard. For a time after his parents left Breslau, he came once a week to her house for lunch.

The Stein children looked forward to his coming. He always had "a wealth of witty ideas" to amuse them. But he also liked to talk to their mother "in his dry and humorous tone" about how he could help his parents. Listening to these conversations, the children did not know "whether to laugh at his funny and extremely exaggerated expressions, or to weep over the content."

In 1904–05, the year after his parents left Breslau, while he was still only in the next to the last year at the gymnasium, Richard began to prepare his girls for the *Abitur*. If they passed this examination, they would be permitted to enter the university. To some of his own teachers, it seemed a reflection upon the importance of the examination that a student who had not yet passed it would presume to prepare others. They tried to discourage his activity and finally told him that he would have to stop tutoring if he wished to continue his studies. He was becoming bored anyway. In the spring of 1905, two weeks after the beginning of his final year, he stopped going to the gymnasium classes and began to attend the lectures in mathematics and physics at the local university.

By this time he had already developed a strong interest in music. He was never able to recall for me the specific origins of it. "I heard music," he said, "and I was impressed." But he did recall vividly an occasion when Julius Stenzel, an older student at the gymnasium and a violinist, substituted for the regular teacher in the music class. At the university he became better acquainted with Stenzel, who was by then an advanced philosophy student; and Stenzel and his younger brother encouraged him to accompany them to chamber music concerts. At some point during this period, he rented a piano and taught himself to play.

(In talking to me, he always objected to the statement that he had taught himself to play the piano. "I never *learned*," he said.)

Stenzel also introduced Richard into a lively circle of young artists, musicians, writers, and students who gathered around Käthe Mugdan, a well-to-do widow.* Stenzel was tutoring her daughter Bertha in the classics and Richard soon became Bertha's tutor in mathematics and physics, and also a close friend of her mother. Many years later, when Käthe Mugdan was over eighty, she committed suicide rather than permit herself to be sent to a concentration camp.

At the end of 1905, Richard applied for permission to take the *Abitur* in the spring with his old class at the König–Wilhelms Gymnasium. He was assigned instead to the class at the Gymnasium zu St. Elisabeth. The difficulty in taking the examination as an outside student can be gauged by the fact that in the group of which he was a member he was the only one to pass.

The following year, 1906–07, Richard continued to attend lectures in mathematics and physics at the university. The difference was that now he had official status as a student: after writing an acceptable dissertation and passing an oral examination, he would be able to obtain a doctor's degree.

He originally wanted to study physics, he told me. It was a time, half a decade after Planck's discovery of the quantum, when the old physics was changing; however, the quantum was never mentioned in lectures in Breslau. Ernst Pringsheim was "a very nice but extremely boring man," and Otto Lummer was "a showman" who amused his students "but didn't let them see what it was all about." To Courant the researches of his

* She was a second cousin of Max Born, and it was in her home that Courant first met Born.

professors seemed "rather divorced" from their teaching. Frustrated by the poor quality of the physics instruction, he turned to mathematics.

Among the mathematicians he found Adolf Kneser "a very original man, one of the moving forces of the time," but an uninspiring lecturer. Rudolf Sturm, from whom he felt that he really learned old-fashioned projective geometry, was simply a drill master. He also heard lectures by Georg Landsberg and Clemens Schaefer, who were not yet professors.

He sat in the classes of these men and thought how much they knew and how little they were able to convey to their students. Of the professors the most successful as a teacher—in his opinion—was by conventional standards the worst lecturer. This was Jacob Rosanes, an algebraist now forgotten in his field. Rosanes came to the platform with a piece of chalk in one hand and a damp sponge in the other. As he lectured, he scribbled equations which his students never quite saw because as he wrote he hid them with his body and as he moved along he rubbed them out with his sponge. In desperation, Richard and other students, including Erich Hecke, stayed after class trying to put together something coherent out of what they had only half heard and scarcely glimpsed. They discovered that in the process they learned a great deal of algebra.

Two older mathematics students, Otto Toeplitz and Ernst Hellinger, with whom Courant had become acquainted the year before at the university, had left Breslau and gone to Göttingen to study. They wrote with enthusiasm of the mathematics at that university and of a mathematician there named Hilbert.

Breslau began to seem unbearably dull to Courant. He felt that it would be good for him to see some mountains, mathematical as well as geographical. His friend Nelly Neumann, who was also attending the university, felt the same way. In the spring of 1907, the two young people went to Zurich to study for a semester.

It was in Zurich that Courant heard the lectures of Adolf Hurwitz, one of the outstanding analysts of the day, the close friend and the real teacher of the mathematician Hilbert, who was so highly praised by Toeplitz and Hellinger. Although to the end of his life Courant always described Hurwitz's lectures as perfect, he was not stirred by them. He returned, still dissatisfied, to Breslau. That fall he took what money he had, bought a new suit, and set out for Göttingen.

TWO

THE EXACT YEAR that he arrived in Göttingen did not remain in Courant's mind. All his life, even in official documents for security clearance, he put it down variously—from as early as 1905 to as late as 1909. He could have ascertained that it was 1907 from the impressive *Diplom* which he brought with him to the United States, but he did not consider the date important. What he never forgot was how he came very soon under the spell of Göttingen and how very young he felt.

In those days, he told me, unless you were met by a horse-drawn carriage, you walked to the town, which was some distance from the railway station. It was little more than a village between gentle hills crowned by the ruins of ancient watch-towers. The newcomer saw first its red-tiled roofs, then came to an old wall and proceeded through narrow, crooked streets to the medieval town hall. In the town square a recently erected statue of a little goose girl, eternally tending her geese at the fountain, was a reminder that the Brothers Grimm had written many of their fairy tales in Göttingen.

Courant arrived at the university in the middle of October, although the semester did not officially begin until November 1. He came early on the advice of Toeplitz and Hellinger, who were very knowledgeable about the scientific situation in Göttingen. During the past year Toeplitz had been the personal assistant of Felix Klein, the organizing force behind mathematics at the university. He was currently a *Privatdozent*, or private lecturer. His mathematical interest, which in Breslau had been algebraic geometry, was now—under the influence of Hilbert—integral equations and the theory of functions of infinitely many variables. Hellinger, who had been Hilbert's assistant during the past year, had just taken his doctor's degree with a dissertation that was to make possible a great advance in this same field. What interested Hilbert—Courant quickly saw—interested almost everybody in Göttingen.

In the opinion of Toeplitz and Hellinger, their little friend from Breslau was going to need all the time he could get to prepare himself for the higher scientific standards of Göttingen. They introduced him to the *Lesezimmer*, the reading room which was the heart of the mathematical life, and put him to work studying electromagnetic theory. If he really tried, he *might* be accepted in Hilbert and Minkowski's joint seminar on mathematical physics.

David Hilbert was by that time the most important mathematician in Germany. He had already produced his classic works on invariants, algebraic number theory, the calculus of variations, and the foundations of geometry. For the past few years he had been working rather exclusively in the field of analysis. Hermann Minkowski, his closest friend since their university days in Königsberg, had made his reputation in a theory of his own creation called the geometry of numbers. When, however, he had joined Hilbert on the faculty in Göttingen, he had suggested that they undertake a study of classical physics for their own education. A result had been the joint seminar on mathematical physics.

Before the semester began, Courant approached Hilbert about being accepted as a member of the seminar and found him agreeable. The topics posted for individual hour-long reports began at a comparatively low level in stationary electrodynamics and proceeded rapidly to a level which Courant recognized as being far beyond his preparation—in spite of the two weeks spent in the *Lesezimmer*. He decided to sign up for the second report, which was scheduled for the third Monday of the semester.

After the semester began, he was very lonely. Toeplitz and Hellinger were busy. The other mathematics students seemed one-sided in their interests, and he saw no potential friends among them. He attended two courses given by Hilbert and one by Minkowski, and found both of the famous mathematicians very bad lecturers. He also took courses in physics and philosophy. But none of the courses interested him very much—only the preparation of his seminar report. He went for long walks by himself in the surrounding hills and decided that in spite of the many charming views he felt "closed in" by the landscape of Göttingen, so different from the plain that surrounded Breslau. It seemed to rain an inordinate amount of the time. When it was raining and he could not walk, he tried to relax from his studies by reading in the cheap attic room he had rented. One of the books he read was *Don Quixote*. His room was already quite cold, and he did not feel he could afford to heat it. Sometimes he thought that perhaps it had been a mistake for him to have come to Göttingen.

He wrote all of these thoughts in long letters to Nelly Neumann, who was still at the university in Breslau. He addressed her as *Liebe Freundin* and signed himself *Ihr Richard Courant*, using throughout the formal *Sie* rather than the intimate *du*. From his letters—I have never seen any of hers—it is apparent that he admired her self-discipline but worried about what he called her "pedantry." She admired his intelligence but warned him against "ambition."

He had hoped that he would find some way to make money in Göttingen; and he was shortly approached by Hilbert himself with the request that he tutor Franz, Hilbert's only son, a strange boy who did not seem to be able to manage the gymnasium course on his own. The job was only temporary—until the boy's regular tutor returned—but in closer contact with Hilbert and observing the social mechanics of the sciences in Göttingen, Courant conceived within the first two weeks of the semester the ambition to become Hilbert's assistant.

The position of assistant to a mathematics professor was unique to Göttingen at that time, although at all German universities it was customary for professors in the experimental sciences to have assistants who cared for laboratory equipment and set up experiments. The salary paid Hilbert's assistant was fifty marks a month and, as was jokingly said in Göttingen, *Familienverkehr*—conversation with Hilbert and his wife, Käthe. That the *Familienverkehr* was more of an attraction than the fifty marks is indicated by the fact that when Courant came to Göttingen, Hilbert's assistant was Alfred Haar, a young man from a wealthy Hungarian family.

The way to the assistantship was through the seminar.

"You are not to take me wrong and think that ambition plays a part in the fact that I am so determined to be in the seminar," Courant hastened to explain to Nelly. "What you write to me about personality and what you consider its most important ingredient—that is also true for me. And referring that to scientific relationships, I have to add that for me personal contact is more valuable than any specifically scientific stimulation. And right now, here in Göttingen—which is like a factory—there is no way to come in contact with the instructors and the professors except by doing something which makes you stand out from the great impersonal mass of students. For inner satisfaction of course there is another way—to go on studying on your own and not bothering about the professors—but that is a very hazardous thing to do. One has to think about the examination, and in my case also the assistantship."

Hilbert had noticed him, been friendly to him. Perhaps he could confirm this apparently favorable opinion when he presented his report in the seminar. But, as he had to report in a long letter to Nelly, "it turned out otherwise."

"To explain what happened, I must first tell you a few things about Hilbert and the seminar," Courant wrote in this letter, dated December 1,

1907. "Hilbert wants, so he says, to acquaint himself with the modern questions in mathematical physics. The individual students are to work through their subjects with the help of the literature so that everything fits into a system of axioms which Hilbert and Minkowski have set up. In doing this, the students are wholly dependent upon themselves, since Hilbert is little informed on the subjects and therefore can hardly give any advice which is useful.

"My report was especially difficult because it was on a beautiful paper by Hertz for which I had to work out a rigorous derivation of the electrostatic and the related electromagnetic phenomena—something which you can't find in books. My predecessor—in fact, the only person who had given a report before me—was a Dr. Helly from Vienna. (Many of the members of the seminar are already doctors, and I am probably the youngest.) He had given his talk on electrostatics to the satisfaction of Hilbert. I had worked so much on my report and had it so well in order that I was fairly certain I would show up well, too. . . . But when I stood in front of the blackboard on Monday, the unexpected occurred.

"I wrote down the fundamental equations from which I had to proceed; and immediately Dr. Helly called out loudly, 'Those equations are wrong!' There was a great commotion, everybody talking very excitedly with one another. Although Hilbert didn't say anything specifically, he looked very surprised, as did Minkowski, who seemed to me rather contemptuous. Although I was completely in the right—and knew it—I was from that moment so confused and intimidated that my report almost slipped out of my mind. I can scarcely remember now what I said. Anyway, I yielded to the objections of Dr. Helly and erased the equations. Then, hardly understanding myself, I again began to talk. But almost immediately Minkowski interrupted me. It seems that in connection with the derivation of the boundary conditions I had failed to take into consideration the surface current—the existence of which I had not been aware of up until that moment. Since the other people in the seminar were not quite clear about it either, it was discussed at length; and my report did not get beyond this point. Hilbert kept getting more and more impatient—although it's surprising that he was not more rude to me—and I became more and more upset and confused. He turned my whole outline upside down, and everybody smiled ironically. The other members of the seminar, especially Helly, talked more than I did. I was happy when the end of the hour came around."

When Courant got back to his attic room, he found a letter from Nelly waiting for him. In it she warned him that it was not good to direct himself singlemindedly toward a goal like the assistantship, and thus make himself dependent upon unknown men and their judgments of him. He agreed eagerly.

"What is important is what a man *is* and not what he appears to be. . . . I have now accustomed myself to the idea that I will not get the assistantship, which will probably become available at Easter. And if it turns out that way, I will begin to think seriously of a career as a teacher, since that would be much better than an unsuccessful life as a not so good scholar!"

The results of the disastrous day in the seminar were not all bad. Suddenly important people began to take notice of him and be very nice, especially Alfred Haar, Hilbert's assistant. Haar personally reproached Dr. Helly for his behavior "and made sure the same thing would not happen again." Toeplitz also came to Courant's support, and Hellinger gave him some good advice on how to improve his report. At the end of the week, Hilbert himself invited him and Dr. Helly and some other members of the seminar to his home on Wilhelm-Weber-Strasse—"which gave me again some courage, although it really doesn't mean anything."

The following Monday Courant presented his report again. This time Dr. Helly did not interrupt at all, and the other members of the seminar held their questions and comments until the end. Hilbert seemed pleased, "especially with some points which I myself had put in order." In fact, everything went so smoothly that Courant could assure Nelly in his letter, written the following day, "the failure of last Monday has been fairly well canceled out."

When he had the seminar report behind him, he found suddenly that in reality all of his courses were very interesting. He got to know another student named Ernst Meissner, who was getting his doctor's degree with Minkowski. "I like him very much. He is indeed—with the exception of Otto Toeplitz—the only mathematician here who is not *only* a mathematician but has also some humanistic interest and understanding." The countryside around Göttingen no longer seemed at all narrow. It had many charms. "The flowing calm line of the hills is very soothing. Everything breathes of peace. Think of it—there is not a single factory here!" He finished *Don Quixote*. "It is marvelous how one never tires of his adventures, although they all have something in common." He found a piano and began to play again, "practicing regularly and trying to concentrate on fundamentals." When Toeplitz told him that he must permit himself heat—it was by now December and really cold—he meekly agreed. He wrote enthusiastically to Nelly, "If only there in Breslau you had someone like Otto Toeplitz, from whose ability to work and reflect one can learn so much!"

Hilbert shortly suggested to Courant that he come to the weekly meetings of the *mathematische Gesellschaft* when a talk that interested him was being given. "This is somewhat unusual, since I am still very young and in general—at the wish of Klein—only students who have the doctor's degree are invited; but Hilbert now rebels everywhere against Klein's assumed dictatorship—and this explains everything."

Minkowski, who by nature was much less outgoing than Hilbert, became friendly too. A few days after the successful seminar report, meeting Courant on the steps of the lecture hall, he stopped and apologized for the rough treatment he had given him. Shortly after, he invited Courant to his house for a special evening with some other students.

Such affairs, I have been told by Minkowski's older daughter, Lily Rüdenberg, were much dreaded by the family. Minkowski was a very shy man, the students were invariably tongue-tied in the presence of a professor, and the evenings dragged painfully. This time, though, things were different. "There was this new, nondescript-looking little fellow from Breslau, and he began asking questions of my father right at the beginning. Pretty soon everybody was talking." It was a marvel in the family and to be remembered all her life by Minkowski's daughter, although she was only a little girl at the time.

My account of Courant's first days in Göttingen is taken for the most part from the letters he wrote to Nelly Neumann at the time. From them it is obvious that in the beginning the most important people in his life were the professors, especially Hilbert. When, however, during his eighty-third year, he talked to me about this period, it was not his success in the seminar that came first to his mind but the fact that he had been almost immediately accepted into what he called the "in group." No one else I talked to about Göttingen ever used this expression, or even seemed particularly impressed by the existence of such a group; but for Courant there was an "in group," and its leader was Alfred Haar.

Haar was a small, delicately built youth with the charming quality of seeming at home anywhere in the world. He possessed the kind of extremely quick, precise talent which was later to be seen at Göttingen in John von Neumann, and was even more knowledgeable than Toeplitz. The nineteen-year-old Courant was very impressed by Haar's savoir faire and his brilliant and witty conversation, also by the material success of his father, who was the owner of great vineyards in Hungary. Even at the very end of his life, whenever he mentioned Haar, he always mentioned the great vineyards of Haar's father. He told me that all the students of his day were convinced that Alfred Haar was the one of their number who would leave the deepest mark on mathematics.

After the success in the seminar, it became Courant's ambition to associate with Haar and his friends. In spite of their comparative wealth, they were not inaccessible to him. German students, even wealthy ones, lived modestly. The chief gathering of the group was the sharing of the hot midday meal at a local boarding house. Such a meal cost about twenty cents in the American money of the time. Although Courant did not feel that he could afford such an expenditure every day, he managed to arrange his budget so that he could dine with Haar and his friends twice a week.

I asked him once how it happened that he had been so quickly accepted. He replied that the "in group" was really very democratic. "When I came, I was friendly with everybody. They needed young people, who were always welcome until they became obnoxious. I guess I must not have become obnoxious."

If he did not offend the members of the "in group," he soon learned that there were others who did. One day, when he and Toeplitz were waiting in the lecture hall for the arrival of Hilbert, Toeplitz began to point out various students and tell Courant something about each of them. At length he came to a big blond young man with glasses, who sat a little apart from the others.

"You see that fellow over there," he said. "That is Mr. Weyl."

From Toeplitz's tone Courant sensed that the big blond young man was for some reason not "in." Toeplitz's next sentence dismissed Hermann Weyl with finality:

"He is someone who also thinks about mathematics."

Weyl was to become, as everyone now knows, the outstanding mathematician of Courant's generation in Göttingen—the true son of Hilbert—a man whose breadth of interest, ranging from the foundations of his subject to its applications in physics, was to equal if not surpass that of his teacher. At the time when Toeplitz made the remark quoted above, Weyl was just completing his doctor's dissertation. He was a country boy, shy and unsophisticated, from Elmshorn, a little town near Hamburg, where his father was an executive clerk in a bank. His offense in his student days, according to Courant and to others who knew him then, was a certain youthful "silliness" at times. He and his intimates conversed in a language in which every syllable was preceded by the sound of the letter p (*peh* in German) ; and once at a party he lay under a chair all evening and answered questions only in barks.

When Courant told me the story about Toeplitz and Weyl, I could see that it still gave him pleasure that in the old student days in Göttingen he had been acceptable to Haar and his friends and Hermann Weyl had not. At the same time it was clear that he appreciated and enjoyed the twist that this fact gave to the whole concept of who was "in" and who was "out."

Alfred Haar became an important mathematician, particularly famous for his very original contributions to the theory of measure; but, as Courant said to me, there was to be no comparison with Weyl.

"No comparison at all."

During that first semester in Göttingen, Courant heard lectures on mathematics, physics, and philosophy.

His interest in philosophy was surprisingly strong, although—as he told me—"it didn't penetrate very lastingly." He took more courses with the phenomenologist Edmund Husserl than with any professor other than Hilbert. The philosophical objective of Husserl was the investigation of the content of human consciousness. He was not concerned with the process of conceptions and thoughts, as the psychologists were, but with the "meaning" of the content and acts of consciousness. He was after the truth upon which all other knowledge rests—"the Archimedean point." He had first applied his phenomenological method to logic and mathematics, and he was always much closer to the mathematicians of Göttingen than to the philosophers. In addition to the lectures of Husserl, Courant also regularly attended philosophical soirees in the rooms of the electric young Leonard Nelson, whose *Habilitation* was being delayed by the opposition of Husserl and the other philosophers but was strongly supported by the mathematicians, especially Hilbert.

Physics was represented in Göttingen during Courant's student days by Woldemar Voigt and Eduard Riecke, both of whom antedated Klein on the faculty. Their presentation of the subject was as classical as the presentation in Breslau had been. Neither was of the stature of Hilbert, Minkowski, and Klein. The impetus which would make Göttingen one of the great physics centers of the world would come from the mathematicians rather than from the physicists. Courant ignored Riecke after the first semester and took only relatively few courses from Voigt, whom he nevertheless admired very much as a person and as a musician.

Mathematics, not physics, was by this time—as he wrote to Nelly— "closest to my heart." He took only two courses with Klein during his five semesters in Göttingen; and although each semester he took a course from Minkowski, he can be said to have "majored" in Hilbert.

For Courant—during his student days and for the rest of his life— there was never to be another teacher to equal Hilbert.

THREE

THE VERY AIR of Göttingen seemed to young Courant to be full of mathematical excitement. Even outside the lecture hall, the students talked intensely "with inspiration and real dedication" about mathematics. This enthusiasm radiated from Hilbert, a slightly built, active man whose independence, originality, and passion for the truth were as apparent in every aspect of his personality and daily life as they were in his great mathematical works.

Seven years before, at the International Mathematical Congress in 1900, Hilbert had listed twenty-three problems, ranging over all of mathematics, on which he had proposed that he and his fellow mathematicians should concentrate their efforts during the coming century. Some of the problems were abstract and general. Others were concrete and individual. Although Hilbert had no interest in what are commonly referred to as the applications of mathematics, he had a great interest in its application to the other sciences. One of his problems had been the mathematization of physics and neighboring subjects. Courant was never to forget the spectacle of a great mathematician "learning physics" at the age of forty-five.

Although on first contact he had found Hilbert a very poor lecturer, "frequently hesitating, varying and amending what he said, sometimes even getting stuck, so that he had to call on his assistant for help," he soon recognized that Hilbert brought to the lecture hall something which even the admired and inspiring Maschke had never been able to bring— a superior mind directly and creatively involved with its subject.

"Hilbert had a way which was much more stimulating than any formal perfection," Courant told me, "because while he lectured he always fought, and the fight was clarifying his ideas. You could follow him. You could *feel* his intellectual muscles."

Like other students of Hilbert, Courant found himself pulled into a struggle to solve a problem, to create or to extend a theory.

Also impressive to Courant was the fact that Hilbert did not isolate himself in order to work. A constant stream of visitors came to the Hilbert house and garden. Discussions of mathematics were interwoven with conversations on a variety of other subjects, especially politics. These were often continued on walks on the wall which still surrounded the town or on hikes into the nearby hills, sometimes on longer expeditions by bicycle. Hilbert ignored distinctions of age and rank. "Whoever you were, you could talk to him as an equal if you had something to say." The Hilbert house was "wide open." Käthe Hilbert was "a wonderful woman," in her own way as independent and original as her husband, full of interest in his young colleagues and students.

Courant was always to consider it his greatest good fortune that, in the fall of 1908, Hilbert—looking around for a student to take Haar's place as assistant—selected Richard Courant, twenty years old.

It was an especially exciting time in Göttingen. Hilbert was attacking the famous theorem of Waring, a problem which had defeated mathematicians for over a century. Minkowski had recently introduced his revolutionary conception of space-time and was working intensely to bring his new ideas to completion.

As Hilbert's assistant, Courant saw the great mathematician several days a week, usually in his garden, where Hilbert always preferred to work. Often Minkowski was present too. In spite of Hilbert's stumbling classroom performance, he spent a great deal of time and thought on his lectures. One of Courant's duties was to research the literature for him and to write reports on relevant points. He also attended Hilbert's lectures, took careful notes, and prepared a neat copy to be placed in the *Lesezimmer*, where the students could consult it at leisure. In addition, he read some of the papers submitted to Hilbert as one of the editors of the *Mathematische Annalen* and helped prepare them for publication. The salary of the assistant was still fifty marks a month (about $12.50 at that time) and *Familienverkehr*.

By Christmas 1908 Hilbert knew that he had proved Waring's theorem. He planned to present his work in the first joint seminar in January, startling even Minkowski, who had been away from Göttingen. But Minkowski was never to hear Hilbert's proof of Waring's theorem. A few days after his return, he suffered a severe attack of appendicitis, was rushed to the hospital and operated upon. He died on January 12, 1909, at the age of forty-five.

(Later in the same year there was another premature death in the Göttingen mathematical circle. Walther Ritz, whose name is attached to the spectral formula and the combination principle of spectral lines, was a *Privatdozent* for theoretical physics. He had been told by his doctors that he was perhaps fatally ill with tuberculosis and should be in a sanatarium under constant medical care. Ritz ignored this advice and worked feverishly on his scientific ideas until his death in July 1909 at the age of thirty-one. A paper by Ritz which made a great impression on Courant was one in which he utilized Hilbert's work on Dirichlet's principle to develop a numerical method of solving boundary-value problems of differential equations. It was fascinating to Courant that a purely theoretical work could lead to a work with practical applications!)

The death of Minkowski turned Hilbert to younger people for the companionship and creative stimulation he had lost. He spent more time with Courant, talked more with him, even gossiped about faculty affairs and the discussion taking place over the choice of a successor to Minkowski. It was from Hilbert that Courant got his first glimpse of Felix Klein as other than the *Bonze*—the important man who takes himself importantly.

The choice for Minkowski's successor, Hilbert told Courant, was between Edmund Landau and Oskar Perron. Both were outstanding mathematicians, but Landau was considered a brassy, rich "Berlin Jew" while Perron was generally well liked. When the time came for the choice, to the surprise of his colleagues, Klein spoke in favor of Landau.

"We being such a group as we are here," Klein said, "it is better if we have a man who is *not easy*."

Klein's opinion was of course decisive, and in the spring of 1909 Landau became a member of the Göttingen faculty.

By the time Landau arrived in Göttingen, Courant had already taken a number of courses dealing with the applications of mathematics, which were well represented at the university as a result of the influence of Klein. He had also attended lectures by Ernst Zermelo on mathematical logic.

(Since Friedrichs had described Courant to me as a mathematician who hated logic, I asked him if he had "hated" the subject even in his student days. "I didn't *hate* logic," he objected. "I was repelled by it. I *believe* in intuition.")

Most important for Courant's future conception of mathematics was the fact that no professor whose lectures he heard in Göttingen was a specialist in one single field. The mathematicians Hilbert, Minkowski, and Klein had a strong interest in physics; the latter two had even considered becoming physicists. Carl Runge, the professor of applied mathematics, had begun his career in pure mathematics. Karl Schwarzschild, the astronomer, made contributions to both theoretical and observational astronomy as well as to quantum mechanics and relativity theory. Ludwig Prandtl, the mechanical genius, developed the mathematical formulation of boundary-layer and airfoil theory. Zermelo, the logician, did his first mathematical work in the calculus of variations and made important contributions to kinetic gas theory.

This general interaction of theory and practice was always to be for Courant the distinguishing characteristic of the Göttingen scientific tradition. Edmund Landau did not fit into this conception of the tradition. His specialty was the analytic theory of numbers, and he loftily dismissed anything connected with the applications of mathematics as *Schmieröl*, or "grease." The new professor lectured as he wrote his many books, proof following theorem without a word of motivation for what he was doing or where he was heading. In Courant's opinion, both lectures and books were sometimes "so abstract that there was no relation at all to substance."

Socially, however, Landau and his wife, Marianne, the daughter of the Nobel laureate Paul Ehrlich, were an addition to life in Göttingen. They gave parties to which students were invited; and Mrs. Landau, like Mrs. Hilbert, took a lively interest in the young people and their problems.

That same spring that Landau assumed Minkowski's chair, Henri Poincaré came to Göttingen to deliver a week of lectures. Hilbert offered his own assistant to the Frenchman, and as a result Courant had the opportunity to observe together the two men who were unanimously acknowledged as the greatest mathematicians in the world at that time. They treated each other with a great deal of respect, he told me, but there was no spark between them like that between Hilbert and Minkowski.

"Which would you rate as the better mathematician?" I asked.

"I am a great admirer of Poincaré," Courant replied. "I think he was the greatest mathematician since Riemann, but you cannot compare him with Hilbert. He did not have the intensity which radiated from Hilbert and which was so wonderful. If he had had that—" More than sixty years later Courant's old eyes brightened at the thought. "But that is unthinkable—Hilbert was absolutely unique!"

During the four semesters that Courant served as Hilbert's assistant, Hilbert was devoting himself almost exclusively to subjects in analysis, the area of mathematics which developed in closest connection with problems of physics and mechanics. Analysis had not been emphasized as a field of advanced endeavor in Breslau. Although the students there studied some of the higher developments of the calculus, they tended to write their dissertations on subjects in algebra and geometry. Nelly confessed to Richard that analysis "frightened" her.

"Don't think that analysis is so much harder than other things," he wrote back. "Perhaps completely the opposite. I am getting into it now, and I find that it is not actually so dangerous."

In fact, Courant took to analysis as if it were his natural element. All his future mathematical work was to be done in subjects to which he was introduced during his student days by Hilbert. It is thus interesting to speculate what might have happened if he had come to Göttingen at a time when Hilbert was working in some other area of mathematics; for Hilbert had the characteristic of becoming so totally immersed in a subject that sometimes he appeared to forget even his own important results in areas of earlier interest.

"I think Courant might then have become the student of Felix Klein," Friedrichs suggested to me.

It is difficult to imagine anyone who would seem to contrast more with Courant than Felix Klein. Courant was small—five feet and five inches in height—and so unprepossessing in appearance that people I interviewed who had known him in his youth could never quite remember what he looked like then. Klein stood out in any group, handsome, dark haired and dark bearded with an impressive head and a memorable smile. He was the acknowledged scientific leader in Germany. Mathematicians, who are not as a rule inclined to hyperbole, invariably recall him as "a Jove." But mathematically Courant and Klein had a great deal in common. Especially notable was the admiration and empathy they had for the work of Bernhard Riemann.

Klein, whose most lasting service to mathematics (in Courant's later opinion) may well have been the opening up of Riemann's work to other mathematicians, arrived in Göttingen, a young Ph.D. from Bonn, just two years after Riemann's death. He was drawn quite early to Riemann's ideas. Many of these had come to that mathematician as the result of a unique geometric intuition which Klein seems to have been almost alone in understanding at the time. Other mathematicians admired the results but were not able to reconcile themselves to the methods by which they had been obtained. Riemann's early death also contributed to the lack of appreciation of his work, but—in Courant's opinion—it is unlikely that even alive he would have had the personal force to put over his ideas. It was left to Klein with his dominating personality and his intuitive comprehension of the geometrical relationships in Riemann's work to become, as Courant later wrote, "the successful apostle of the Riemannian spirit."

"The question, however, is *Why was Courant not immediately attracted to Klein?*" Friedrichs mused. "For Hilbert's way was not natural to Courant, as Klein's was. Of course Hilbert was *the* mathematician in Göttingen—at that time everyone gathered around Hilbert—but I think the fact that Courant was so very attracted to Hilbert has something to do with the way he was often drawn to people who were personally and mathematically completely different from him. He always had unbounded admiration for such people. In regard to Klein he felt, I think—'that I can do myself; but Hilbert, now there is something really different!' "

From the time of Courant's arrival in Göttingen, he expected to get his degree with Hilbert.

"But one must think of the drawbacks in working with Hilbert . . . ," he wrote judiciously to Nelly in his first weeks at the university. "He doesn't bother very much about the people who work with him, and people other than his assistants rarely have the privilege of receiving direct stimulation from him. Everybody says he hands out doctoral topics without regard to the personality and preparation of the students in question so that out of ten dissertations seven go completely wrong and of the other three, generally two are discovered after promotion to have been incorrect. However, since I always have firm mathematical support here in Toeplitz and also Hellinger, I don't think the danger is too great that I will get myself involved in an unfruitful topic."

As it happened, the topic which Hilbert did suggest to Courant for his dissertation turned out to be so remarkably suited to Courant's personality and to his mode of mathematical thought that it was to run, in the German expression, "like a red thread" through his life's work. This was *Dirichlet's principle.*

I asked Friedrichs why he thought that Dirichlet's principle had such a powerful attraction for Courant.

"It has esthetic appeal," he replied promptly. "There is a solution of a problem that is characterized by the fact that a certain integral, which is a very simple and clear-cut one, assumes a minimum for that solution. This fascinated him. He found it so convincing and so beautiful."

Already—in Courant's student days—Dirichlet's principle had more than half a century of high intellectual drama behind it; and for the rest of his life Courant was always to enjoy recounting its history in talks and papers meant for general audiences. He even devoted a section to it in his book *What Is Mathematics?*, causing the American mathematician E. T. Bell to remark that it was probably the first time in the history of mathematics that Dirichlet's principle had appeared in an elementary mathematical book.

The story of Dirichlet's principle, as Courant always told it, began with Gauss, who in his proof of the fundamental theorem of algebra made it clear that the first step in solving a mathematical equation is to establish that a solution of the equation does in fact exist. In the case of an algebraic equation, this step is relatively simple; but in the case of a differential equation, it can be a very knotty problem. It was for the solving of such a problem that the mode of reasoning later to be identified as Dirichlet's principle was first employed.

The problem concerned a differential equation, known as the Laplace equation, which is of fundamental importance in algebraic geometry and in mathematical physics. The solutions of this equation characterize the so-called harmonic surfaces, and the problem is to determine a harmonic surface that is bounded by a given contour. But not only is it difficult to determine the desired surface—it is also not at all certain that such a surface exists.

During the time of Gauss, however, it was observed that there is a certain integral which can be formed for any sufficiently smooth surface. This integral has the property that its value for a harmonic surface bounded by a given contour is smaller than for any other such surface. Conversely, any surface that has the given boundary and minimizes the integral is harmonic. It seemed clear that because of the positive character of the integral it would assume a minimum; and thus the existence of the desired surface—and hence a solution of the equation—was considered to be established.

In the lectures of Dirichlet, who succeeded Gauss in Göttingen, the young Riemann became acquainted with this reasoning. Impressed—as Courant said—"by its elegant simplicity and what seemed to him its inherent truth," Riemann utilized the same reasoning for the proof of a number of the fundamental theorems of geometric function theory and the theory of Abelian functions. And he gave it the name *Dirichlet's principle.*

Some time after the publication of Riemann's works, Weierstrass, one of the leading German mathematicians of the day, objected to Riemann's assumption that the "Dirichlet integral" necessarily assumes a minimum. Weierstrass's objection was the following. Since the positive character of the integral implies the existence of a lower bound to its values, Riemann was taking it for granted that this bound was a proper minimum—in other words, a minimum that would in fact be assumed by one of the admissible surfaces, or functions. But, Weierstrass pointed out, while continuous functions of a finite number of variables always possess a least value in a closed domain, there may be minimum problems of a more involved kind for which the relevant integral does not assume a minimum —even though a lower bound exists.

Riemann recognized the validity of Weierstrass's criticism. Still, he was intuitively convinced of the truth of Dirichlet's principle and of the truth of his theorems, the proof of which rested on the postulation of the principle. He died in 1866 at the age of forty without having answered Weierstrass's criticism. Shortly afterwards, Weierstrass was able to produce a very delicate and special example of a minimum problem for which the integral does not assume a minimum in spite of having a greatest lower bound.

By this time many mathematicians felt that they simply could not give up the results which had been gained by Riemann's methods. They tried to shore up Dirichlet's principle as best they could; and when they were unsuccessful, they invented various ad hoc ways of proving the important existence theorems which Riemann had thought to be derived so simply by means of the principle.

This was the situation when—almost half a century after the original work of Riemann—Hilbert "rehabilitated" Dirichlet's principle. He did this by proving *directly* that the minimum problem in question does in fact have a solution *by first finding the appropriate surface and then verifying that for it the integral assumes a minimum.*

For Courant this direct approach to the problem was always to be among the deepest and most powerful of Hilbert's mathematical achievements.

"To attack what seemed the absolutely inaccessible problem of salvaging the Dirichlet principle required the complete lack of bias and the freedom from the pressure of tradition which are characteristic of truly great scientists," he wrote on Hilbert's sixtieth birthday in 1922. "Hilbert possessed the courage, he tried and he succeeded. Completely novel and sophisticated methods of the most refined and penetrating insight were applied, and the reader of the work had to struggle laboriously to comprehend. But the great final goal had been achieved. A new weapon had been shaped by means of which an actual method of proof could be created out of the Dirichlet principle."

It happened that in 1909, when Courant was looking for a topic for his dissertation, there was a great deal of interest among the young people around Hilbert in the uniformization of Riemann surfaces and in the conformal mapping of such a surface onto a kind of standard domain. An important breakthrough in these problems had recently been made by Paul Koebe, a somewhat older man who was a *Privatdozent* in Göttingen at the time. By proving the uniformization theorem, Koebe had established conditions under which certain Riemann surfaces could be mapped one-to-one conformally on certain other domains. Hilbert, who had recently become interested in this general area, suggested to Courant that in his dissertation he should try to apply Dirichlet's principle to the same problems.

FOUR

OURANT'S dissertation was entitled "On the application of Dirichlet's principle to the problems of conformal mapping." In it he was able to modify and simplify Hilbert's approach to the principle and to apply it to certain fundamental problems of geometric function theory concerning, among other things, the conformal mapping of Riemann surfaces of higher genus. He was able to prove in a new and different way the uniformization theorem which Koebe had proved in 1908. He was also able to prove another important theorem, known as the slit theorem, which had been stated very generally in 1909 by Hilbert. He developed as well an estimate which later led him to the statement of a lemma that was to become one of his most powerful and most frequently used tools.

Unfortunately, even before he had finished writing this maiden paper, Courant's satisfaction in it was marred in a disagreeable way.

In the published dissertation (1910), there are a number of references to Koebe and his work in the footnotes. One of these is unusual.

"While I was writing this paper," Courant noted on page 5, "there appeared . . . a paper by Mr. Koebe, the essential content of which is the proof of this same theorem, which was first formulated by Hilbert as a general theorem. Mr. Koebe's proof is also based on Hilbert's idea of employing the minimum property of the flow potential and agrees further with my proof of the convergence of these potentials. Therefore, I would like to emphasize here that I have found my proof myself and independently of Mr. Koebe and have so informed Geheimrat Hilbert of this in June 1909."

I asked Courant for the story behind the footnote.

"It was one of those unpleasant experiences," he said. "Koebe was already well established, and very knowing, and he could not stand having someone else do something in his field. One day he asked me—I remember it very well—what I had done in my thesis. He asked me on the street in Göttingen. We had quite a walk and I told him what I had done and then he said, 'Is that all?' I was somewhat taken aback. Then in a few days he came to me and said he had done the same thing in another way, which was more closely connected to what he had done before."

The following week, when Courant was scheduled to talk on his work before the *mathematische Gesellschaft*, Koebe announced at the beginning of the meeting that he would also like to talk on the same subject.

"Well, we have two gentlemen who wish to talk," said Klein, who was the chairman. "Who will talk first?"

Koebe promptly rose to his feet and said, "I will."

Courant could not object, since Koebe was senior to him. The other members of the *Gesellschaft*, however, were very disapproving of Koebe's behavior. Courant was six years younger and a beginner. He had written a very nice dissertation, and he should have had the privilege of presenting his results first.

Some of Courant's friends—one of them was Kurt Hahn, who later became internationally known as the founder of famous schools in Germany and Scotland—decided to pay Koebe back for his treatment of Courant. They rigged up a sophisticated mechanism which would set off an alarm at irregular intervals and hid this in a chamber pot under the lectern before one of Koebe's lectures. There was much laughter at Koebe's expense when he finally located the offending mechanism and drew it out in its container. Hahn then further publicized the prank with an article in a local newspaper.

I have been told by people other than Courant that Koebe was considered a conceited and disagreeable man with a reputation for picking up the ideas of younger people and, because he was so quick, being able to finalize and publish them first. He was, nevertheless, an outstanding mathematician; and at the end of the summer semester 1909 he was called to a professorship at another university and left Göttingen.

After Hilbert approved the dissertation, nothing stood between Courant and the coveted degree but the oral examination. This was set for six o'clock in the evening on February 16, 1910. The three examiners were Hilbert, for mathematics; Voigt, for physics; and Husserl, for philosophy.

Hilbert arrived on time and before the others. He was eager to get on with the examination and go home. But Voigt and Husserl did not appear.

"Well, well," Hilbert said at last to Courant, "I could of course start now and ask you some questions; but I know you so well, talking to you about mathematics two or three times a week, why should I ask you questions? Of course I could do that. Ja. But it would have no point."

So he sat and chatted with Courant about various non-mathematical subjects, looked frequently at his watch, wished aloud that Voigt or Husserl would come—"so that they could ask you some questions"—and then chatted some more about subjects of mutual interest and people they knew.

At last, forty minutes after the appointed time, Husserl arrived.

"Oh, Husserl," Hilbert said with a great show of relief, "now you are here so I can stop asking Mr. Courant questions."

He promptly excused himself and went home to dinner.

Husserl asked Courant one question. Before he could ask another, Courant ask him to explain a delicate point of phenomenology. Husserl talked until the time allotted for the examination was up. Voigt never did arrive.

After the oral examination and the granting of the Ph.D., Kurt Hahn decided to bring Courant's new status to the attention of the community. He and two other students rented a small horse-drawn carriage and equipped themselves with megaphones. Then, with the embarrassed Courant shrunk down in the back seat to make himself as nearly invisible as possible, they proceeded to drive around the town, trumpeting to the inhabitants of Göttingen through the quiet night, "Dr. Richard Courant *summa cum laude!*"

The next professional step was the *Habilitation*, which carried with it the license to deliver lectures as a *Privatdozent* and, although not paid a salary by the university, to collect fees from students. The decision to take this step did not lie with the young doctor but with the faculty. Klein or Hilbert took one aside—perhaps on the annual picnic and hike for mathematics faculty and students—and suggested that one might like to stay in Göttingen.

The number of private lecturers was carefully controlled. When Courant received his degree, there were seven for mathematics. Until some of these moved on to positions at other universities, a job as assistant to Hilbert looked very good indeed to Courant; and he decided to hold onto it.

There were beginning to be signs that Hilbert was moving into a new area of research. During the winter semester 1909–10, he had lectured exclusively on differential equations, the principal mathematcal tools of physics; and during the summer semester 1910, he continued to concentrate on these in his lectures.

As assistant to Hilbert, Courant helped as in the past with the preparation of the lectures and took notes during their delivery. He produced hardly any mathematical work.

"I was very often depressed," he said to me, explaining a rather melancholy picture of himself taken at the time, and added, because he was by then many times a grandfather, "We forget that these things are not new."

During the late summer of 1910, he spent some time in Berlin at the home of his friend Kurt Hahn. His own parents were not in the capital. According to Siegmund's memoirs, they had been "exiled" by a younger brother, Eugen, and lived in another town with distant relatives who were under orders from Eugen to give them "one mark a day" to live on.

Courant had recently received orders to report for the compulsory year of military service. The friendship between him and Nelly Neumann had ripened into what both of them thought was love. He hoped that when he finished his army duty, he would become a *Privatdozent* in Göttingen. Then he and Nelly would be married.

On October 2, 1910, on the eve of his induction into the army, he decided to begin a diary, both as a record of his life "for Nelly and our children" and as a tool for self-examination and self-improvement.

"The decisive experiences of my life are behind me," he wrote in his student's notebook. "The course which it will take is in all probability determined within definite boundaries...."

When he looked back over the past six or seven years, he found that he was content "with what fate has bestowed on me in love, in the friendship of people, and in opportunities. And if I am still not a happy and satisfied man—if I still think that I have not made of myself what I could have made—then the reason is not the fact that I have not been spared bitter experiences and worse circumstances than most people have ever known. It is me alone.

"During these last important years I have trotted through the world as if with blinders on. I have become a scholar who does not understand much about his chosen science—who does not possess the disdain of many scholarly souls for all that is not known to them, but who still truly knows little about a thousand things about which he should know much...."

He wasted too much time talking with people. "And I will always place the human far above the intellectual!" But he hoped that keeping a journal would help him achieve the self-discipline he felt he lacked and would also help him improve his "frighteningly bad" memory for facts and experiences, "which deeply disturbs me."

It was comforting that Nelly knew his weaknesses and that they did not affect her love for him. "But no one wishes more than she that things get better."

The moment of induction into the army seemed to him propitious for such bettering, more propitious than those times in the past when he had earlier made such resolutions. "This time from outside an actual break has been made in my life. From tomorrow on, strict discipline will be applied to me. Now I must pull myself together so that it is also thus within me."

Perhaps to his own surprise, Courant found that he was more suited to the soldier's life than a young scholar might reasonably have expected himself to be. He had no difficulty in relating to the other men, although most of them were not university people but farmers from the countryside surrounding Göttingen. During his year of service he became a non-commissioned officer. But the military life was not moving him ahead in his academic career. He was envious of fellow students who did not have to share the burden. He visited Göttingen whenever he had the opportunity. He hoped to be asked to return for *Habilitation,* but there was no suggestion to that effect from either Klein or Hilbert. Then in May 1911, while he was still on duty with the reserve, he received a letter from Kurt Hensel, who was a professor at Marburg, proposing that he come there and, while preparing the thesis required for *Habilitation,* assist Hensel with the editing of the collected works of Kronecker.

Courant put off replying to Hensel's letter until, on leave, he could communicate the Marburg proposal in Göttingen. When after three weeks he still had not given Hensel an answer, he received a letter from Hellinger, then a *Privatdozent* in Marburg, scolding him for his delay. By this time Hilbert had suggested that Courant return to Göttingen. Courant wrote promptly to Hellinger and enclosed a copy of the letter he planned to send to Hensel.

"The idea of *Habilitation* in Marburg has always been tempting to me," Courant had written in his letter to Hensel, "and, therefore, your suggestion in this connection coincided with my own wish. I had looked forward very much to the activities in Marburg and particularly to the cooperation with you in connection with the editing of Kronecker's works. But suddenly now there has been a chance to remain in Göttingen, where I can get my *Habilitation* without difficulty during the winter semester."

Hellinger firmly noted on the letter: "To be changed—it sounds as if you were going to Marburg for lack of someplace better to go."

In the next paragraph Courant had spelled out the attractions of Göttingen.

"Geheimrat Hilbert promises me that at least once a year I will get to give one of the main big four-hour lectures. Because of the greater number of students in Göttingen, which will presumably not decline in the coming years, the financial side of the matter is very favorable. Even though I will have to do with considerably less income than in the old Göttingen three-lecturer arrangement, my annual income will nevertheless be considerably higher than what I could expect it to be in Marburg. Apart from my many scientific and personal connections with Göttingen—particularly with my highly esteemed teacher Hilbert—the financial consideration has been the deciding factor. I would regret it very much if this decison of mine would cause an unwelcome disturbance in your disposition."

Hellinger wrote disgustedly: "Express yourself less patronizingly."

"In any case," Courant had continued, "I ask you not to hold my decision against me. After considerable consultation with my friends and with Hilbert, I have come to the firm conclusion that I should not turn down the offer of *Habilitation* in Göttingen."

Hellinger crossed out "my friends" and wrote, "Leave them out of this."

When Hensel received Courant's letter, he replied cordially, "I can understand why you would wish to remain close to Mr. Hilbert."

In the fall of 1911, his year of military service behind him, the prospect of *Habilitation* in Göttingen ahead, Courant became engaged to Nelly Neumann.

Nelly had received her doctor's degree in Breslau the year before Courant took his in Göttingen. Her subject was a topic in geometry. Afterwards she had taken the examination and written the additional thesis required of prospective gymnasium teachers. Although "talkative and gay," she had grown into a woman who was "very thorough," according to Edith Stein. She was "particularly interested in ethical questions" and "never undertook to do anything without having carefully weighed all the pros and cons."

Courant felt that by bringing himself up to Nelly's high standards he would become the person he wanted to be. On the other hand, although she was two years older than he, he felt very protective toward her. He worried about her living habits and her health. She was inclined to drive herself. One of his letters to her ends with the exhortation: "Don't forget to swim, do gymnastics, work in the garden, take walks!"

In the journal which he had begun on the eve of his induction into the army, he had written of their relationship:

"Nelly knows every smallest and most secret thing about me, and it is wonderful and reassuring that I need to conceal nothing from her. Even though she disapproves of much, that does not affect her love. It is completely independent of such things."

The engagement was not yet formally announced, and he told nobody in Göttingen about it except Mrs. Hilbert and Mrs. Landau.

"Mrs. Hilbert said that as soon as it is official, you must come to visit me. Let's hope that will be at Christmas!"

The time between his discharge and his *Habilitation* was again an economically difficult one for Courant. During his military service he had regularly sent 120 marks a month to his parents and this had enabled them to return to Berlin. Now, no longer able to help them himself, he negotiated an allowance for them from the family business in Lublinitz. Part of the agreement was that the money was to go through him and that "every penny" was to be accounted for.

The position of Hilbert's assistant was currently being held by Erich Hecke, whom Courant had recommended for the job before leaving for the army; but Hilbert needed some additional help with his book on the foundations of a general theory of integral equations. Courant read and corrected proofs. He also found a couple of mathematics students to tutor.

"My dear, dear girl," he wrote to Nelly on the eve of the new semester, ". . . you can forgive and forget my transgressions against you, for I love you far better than it seems at times. . . . I am going to really try to conquer all the faults which you reprove me for, and with your help I will be successful."

This letter to Nelly is the only one of the period that Courant kept. The other letters I have seen are all from the first few months when he came to Göttingen. He may have kept these particular letters because they contained descriptions of people and events he considered of historical interest. As he told me in our first conversation, he had always intended to write something about the scientific life in Göttingen. This letter, dated October 27, 1911, in addition to describing a busy, bustling trip to Berlin, reports on events in the philosophical circles of Göttingen.

At this time Courant was dining once a week with Leonard Nelson, who was at last a *Privatdozent*. Nelson attracted many young people and was once described by an admirer as coming nearer to Socrates than any other modern. He wanted to establish by rigorous reasoning a system of philosophy in which everything had its determined place and every essential question was answered. He spoke of a "philosophy of right" and tried to lead his followers to a clarification and a critical examination of their own convictions. He then required them to carry out in their actions what they had recognized as just and good.

"Ethics," he told them, "is there to be applied."

Philosophically, Courant found Nelson more of "a medieval dogmatist" than a Socrates.

"Were you ever a follower of his?"

"No," he replied. "I could not stand his philosophy. I was a friend. You could not be a friend *and* a follower."

Courant's letter to Nelly also describes the problems attendant upon his proposed *Habilitation*. Now that he was back in Göttingen, the arrangements for that step did not seem to be nearly so certain as Hilbert had implied.

"Hilbert has talked with other people here, and there may be difficulties because some of them think I should perhaps have written more papers. Still, according to Hilbert, everybody wants me to be ready to start to lecture at Easter. Hilbert put it this way: 'We hope that you will have produced by then a beautiful *Habilitation* thesis.' Well. We shall see."

The winter of 1911–12, when Courant was preparing his *Habilitation* thesis, was marked by a high-spirited celebration of Hilbert's fiftieth birthday on January 23. Assistants and advanced students composed a number of verses for the occasion. One set of these described the "creation" by Klein of the position of Hilbert's assistant and the many duties which were performed for 50 marks a month and *Familienverkehr*. It concluded with the creation by Hilbert's current assistant, Hecke—"a practical man"—of the position of "assistant to the assistant." This luck-less fellow did all the work but did not receive the 50 marks: "His pay for this is—ach, how nice—/ *Familienverkehr*/ And, besides, Hilbert as his boss—/ O heart, what wouldst thou more!"

The highlight of the celebration was a tour de force which devoted a couplet to each of Hilbert's romantic attachments—usually the young wife of some colleague. Beginning with Ada, Bertha and Clara, it proceeded triumphantly through the alphabet, stumbling only over Q, U, X, Y, and Z. Mrs. Hilbert was not forgotten: "God be thanked that Käthe, his wife/ Takes not too seriously his life."

(When I was writing *Hilbert*, I questioned Courant about the nature of Hilbert's relationships with his many "flames," as they were called. One mathematician had insisted to me that they were indeed affairs. "That man!" Courant said disgustedly. "He is no gentleman. He doesn't *under-stand* such things!")

For his *Habilitation* thesis, Courant returned to Dirichlet's principle. In a paper entitled "On the method of the Dirichlet principle," he gave a proof of the validity of the principle by means of a method which was close to Hilbert's second method but avoided various complications which were due to the great generality of Hilbert's work.

The next step in *Habilitation* was the delivery of an inaugural lecture before the Philosophical Faculty. According to custom, Courant submitted three topics of his own choice. By the time I talked to him, he had forgotten two of the topics; but he remembered the one approved by the faculty— "On existence proofs in mathematics." This was the subject on which he had personally most wanted to talk.

The interest in existence proofs was to run, with Dirichlet's principle, all through his work.

"Even the book I am trying to do now," he told me, referring to a third volume of Courant-Hilbert which was never to be completed, "is in a way about existence proofs."

Courant's inaugural lecture was delivered (without notes, as was the custom) on February 23, 1912. The dean of the faculty wrote to him on the same day, congratulating him and announcing that he had been granted the license to lecture in Göttingen.

At Easter, when the new semester began, he delivered his first lec-tures; and he and Nelly Neumann set a date for their marriage.

FIVE

WHEN, after the long summer vacation of 1912, Dr. Courant returned to Göttingen with a wife, it was an occasion for some surprise, especially since she was two years older than he, and rich.

"We always had the impression that their only common interest was mathematics," explained his former pupil Magda Frei, who was by that time also a young wife in Göttingen. "Nelly was the best in our class in mathematics, but not the best in everything. She never seemed to me like the marrying kind. It was a little—well, I thought at the time he probably felt 'obliged.' There was the question, could he live on what he made tutoring. I always thought her father must have helped him."

The newly married couple rented an apartment on Schillerstrasse, and Courant began to try to establish there the kind of "wide open" academic home to which he had been introduced in Göttingen. He had always admired the way in which Mrs. Husserl entertained, her housekeeping skills and cooking ability as well as the conversation at her table. The philosopher was still not a full professor, and his family was less well off than most. All was done frugally but with charm. One of the first people Courant invited to meet his new wife was the Husserls' daughter Elli, a student of art history at the university.

"I still remember it," she told me over tea at her home near Cambridge. "They lived a little outside the town. That was a time when, I think, there wasn't a single bus in Göttingen, so I walked there. And the tea—it was just very nice, but it wasn't the kind of lively atmosphere to which I was accustomed. He obviously tried his best. I remember only that one visit. It was not a roaring success. It must have been very hard for her."

In spite of Richard's efforts, Nelly was lonely. She suggested to Aunt Auguste Stein that she send her daughter Edith, who was already interested in philosophy, to study with Husserl in Göttingen. "Richard has many friends whom he has brought into our marriage, but few women friends. This would be some compensation for me." She repeated the suggestion when she was in Breslau a few months after the wedding. She also described her first adventures in housekeeping. "Things become increasingly more complicated the farther they are removed from mathematics," she observed, "and the household is the farthest removed."

As a result of Nelly's urging, Aunt Auguste permitted Edith and her older sister Rose to go to Göttingen in the spring of 1913. They were immediately made welcome by the young Courants. "Without wanting it to be true, Richard was still greatly attached to the family," Edith noted, "and always asked me about every member."

Unlike the more sophisticated Elli Husserl, Edith found the apartment in Schillerstrasse a place of easy hospitality: "Richard loved to bring unannounced guests. He had a great circle of friends, other instructors and older students; but he also liked to bring home some of his own students, young men and women with whom he wanted to discuss matters."

To Edith it seemed that "his wife's fortune" had provided her cousin Richard for the first time with "an existence free from worry and full of the youthful unstinted joy of life."

During the years from 1912 to 1914 when Courant was a young teacher in Göttingen, the famous faculty of his student days was beginning to change.

For most of 1912, Felix Klein was in a sanatarium in the nearby Harz Mountains. He continued to keep in touch with events at the university and to direct them; but at the end of the year, although he was only sixty-three, he wrote to the ministry and asked to be made emeritus.

In 1913 Constantin Carathéodory succeeded to the chair of Klein. Carathéodory was one of the many young men destined to become important scientists who received their degrees in Göttingen between 1900 and 1910. Among the others, in order of time, were Max Dehn, Felix Bernstein, Georg Hamel, Rudolf Fueter, Oliver Kellogg, Erhard Schmidt, Max Born, Ernst Hellinger, Alfred Haar, Hermann Weyl, Richard Courant, and Erich Hecke. If Klein, passing his sixtieth birthday in 1909—at the time of Poincaré's visit—had begun to look among these younger men for the one who would bring to reality his own dream of a great mathematics institute in Göttingen, it is unlikely that he would have perceived him in Courant. Most probably Klein's selection would have been Carathéodory, the scion of a powerful and wealthy Greek family, a man of unusual culture and sophistication, somewhat older than his fellow students since he had been an engineer before he decided to become a mathematician; one who in his doctoral dissertation in the calculus of variations had shown exceptional power and originality; a tall, impressive, aristocratic-looking man, although unfortunately cross-eyed. As a further recommendation for Klein, Carathéodory had an interest in educational problems. He had attended the École Polytechnique in Brussels, founded in the spirit of the École Polytechnique in Paris—a school which had revolutionized scientific education in a way which Klein much admired. In short, Carathéodory was a natural to occupy Klein's chair and his return to Göttingen was universally approved.

The scientific atmosphere of the university was also changing during Courant's days as a young teacher.

In 1912 Hilbert turned his attention from mathematics to physics with his well-known pronouncement that physics was much too hard for physicists. He asked his friend Arnold Sommerfeld in Munich to send an advanced student to Göttingen to be his physics assistant. The first such young man sent by Sommerfeld—there was to be a series that extended through the 1920's—was Paul Ewald.

Ewald and his wife, Ella, were among the few people I was able to interview who had known Courant in his youth. They remembered him as being "serious" but, as Ewald said, "having the wonderful gift of making fun of himself—this typically Jewish attitude of considering oneself as a little poor boy up against the big people on top." There was also a certain "eastern" Jewishness about him, although he was not really an eastern Jew, which—in Ewald's opinion—charmed the western "integrated" Jews of Göttingen.

After Ewald became "Hilbert's tutor for physics" (as he was promptly nicknamed), Hilbert began to look for an outstanding physicist for the faculty. He found such a person in a young Dutchman named Peter Debye, who had been Sommerfeld's first assistant in Munich; and he personally negotiated Debye's appointment to a professorship in Göttingen.

With Hilbert's new interest in physics, important physicists began to be invited to report on the latest discoveries by themselves and their students. One of the first of these was Sommerfeld himself, who came in 1912 to tell about the work of von Laue and others on X-ray diffraction in crystals, a work made possible by mathematical formulations in Ewald's not yet completed doctoral dissertation.

It may have been on this occasion that Sommerfeld talked to Courant about the quality of books on mathematical physics. He said that the older generation had produced some good books but that now something new was needed. Since his own student days, Courant had been aware of the deficiencies of most of the textbooks available, both in mathematics and mathematical physics. He had thought seriously that some day "when he grew up" he might do something about them. He did not forget Sommerfeld's remarks.

During this period Courant's fellow instructors in Göttingen included Max Born, for theoretical physics; Toeplitz, Weyl and Hecke, for mathematics; and Theodor Kármán, for mechanics (the *von*, "the sign of nobility," as Courant always liked to explain, came later).

Born was from Breslau. Like Courant, he had come to "the mecca of mathematics" on the recommendation of Toeplitz and Hellinger, but even before they themselves had come. He had been Hilbert's assistant in 1905–06 and had taken his degree in 1907. After Minkowski's death he had been asked by Hilbert to edit Minkowski's physics works.

Kármán had been introduced into the Göttingen circle by his countryman Alfred Haar, who had since become a professor in Hungary. With his vitality and charm, he had soon succeeded to Haar's position as the leader of what Courant still saw as the "in group." One of the most memorable non-scientific events of the period, at least for Courant, occurred when Kármán and the "outsider" Hermann Weyl, having both fallen in love with the beautiful and gifted Hella Joseph, got into a water fight over her. Hella had to make up her mind on the spot whose face she would wipe and, much to the chagrin of the "in group," she chose to wipe Weyl's.

Another foreign student who was an integral part of the Göttingen scientific community was Harald Bohr. At that time he was the better known of the Bohr brothers, having been a member of the Danish soccer team which had been the runner-up in the Olympics of 1908. He was also a promising mathematician and had come to Göttingen at the invitation of Landau. Everyone was attracted by his intelligence, worldliness, wide interests, and "radiance" (Courant's word); but Harald Bohr insisted that he was nothing in comparison with his older brother, "who was made of pure gold and would soon be recognized as one of the great scientists of our time." This excessive admiration for his brother was looked upon by most people in Göttingen as a foolish but endearingly modest trait in the otherwise wholly admirable Bohr. Courant, accepting Harald almost worshipfully, also accepted his faith in Niels Bohr.

All these young men were beginning to do work that attracted notice in the scientific world. Weyl published his lectures on Riemann surfaces. Hecke's dissertation on number theory was a significant piece of work and was followed by other important papers. Born and Kármán, rooming together in the legendary ElBoKaReBo (its name derived from the first syllables of the names of its five inhabitants), collaborated on a work in which they stated, independently of Einstein, the laws of specific heat. What Courant recalled sixty years later, however, was the impression made in Göttingen by two works from the "outside" world.

The first of these was Niels Bohr's revolutionary interpretation of the atom, which appeared in 1913—what has been called (by James Franck) "the birth hour of atomic physics." On a visit to Cambridge that same year, Courant had met "Harald's brother" and had heard him set forth his ideas in the quadrangle of Trinity.

"Thanks to prior suggestion by Harald, who had so often told me wonderful things about his brother, I was at that point immediately ready to believe that you must be right," he wrote later to Niels Bohr. "But when I then reported these things here in Göttingen, they laughed at me that I should take such fantasies seriously. Thus I became, so to speak, a martyr to the Bohr model. . . ."

The other work from the outside was that of a young American named George Birkhoff. According to Courant, when Poincaré had lectured in Göttingen, he had mentioned a certain theorem which he had been trying without success to establish. Later, although he still had not been able to prove it, he had published his work on the theorem in the hope that his results might at least be useful to others. After Poincaré's death in the summer of 1912, the theorem became "Poincaré's last theorem"—a challenge to mathematicians all over the world. In Göttingen the young men said, "That doesn't seem too hard—some rainy afternoon when we have nothing else to do we shall think about it and prove it." But the rains came and went, and even in Göttingen the last theorem of Henri Poincaré turned out to be as hard a nut to crack as it had been in Paris. Then, in the fall of 1913, a paper proving the theorem was reviewed in the *Fortschritte der Mathematik*. It had been published in what Courant always referred to as "some obscure American journal." It seems that all American journals were considered obscure in Göttingen. Overnight everyone was talking excitedly about the author, *der Amerikaner Birkhoff*.

Many Americans had studied in Göttingen; and although they were well liked and sometimes did mathematical things which people admired, they were not considered quite "top drawer" as mathematicians. In 1913, for the first time in Courant's memory, the Göttingers looked with admiration across the ocean.

In Göttingen during this period, the general direction of the mathematics and physics activities remained still in the hands of Klein. As an emeritus professor he was relieved of "duties" but not of "rights and privileges." He continued to be—as he had been for more than a quarter of a century—the strongest directing force in the faculty, the most influential academic voice at the ministry in Berlin, and the representative of the university in relation to industry.

In a few years the *Vereinigung zur Förderung der Angewandten Physik und Mathematik,* the association of businessmen and industrialists which Klein had organized to support science at the university, was to celebrate its twentieth anniversary. In 1913, in preparation for a history of the association which was to be published at that time, Klein went back over his professional life, year by year. His notes, based on letters he had written at the time, are among his papers in Göttingen. In them one can glimpse the exertions, the developing strains, the illnesses which led to the breakdown that effectively ended his career as a creative mathematician at the age of thirty-three. It is clear from these notes that he recognized immediately the effect that his illness would have. "From now on," he wrote, "social effectiveness will have to substitute for lost genius." The notes also show how early the outlines for the development in Göttingen began to form in Klein's mind.

In 1914 the idea for an institute of mathematics in close physical and intellectual contact with the applications—which was to be the climax of Klein's endeavors—was on the verge of being realized. Land next to the physics institute had been given to the university, plans for the building had been drawn up, construction was ready to begin.

In later years Courant was always to recall the pre-war period in Göttingen as idyllic. The whole mathematics group—faculty and students —was "somewhat like a family." The university was "an ideal place for young scholars searching for eternal values without taking themselves too seriously"—the struggle "tempered by healthy enjoyment of an unpretentious but meaningful social life, based on real human contacts and the beautiful surroundings of the town." But in spite of the generally happy situation, Courant could not help seeing that with the appointment of Carathéodory to Klein's chair, opportunities for advancement to a professorship in Göttingen were effectively closed off. The chair of Runge, the oldest of the other three mathematics professors, would not be vacant until 1925. Hilbert would not retire until 1930, and Landau not until 1945. It was clear to Courant that he and his contemporaries would have to look to other universities for their professorships.

Courant had produced relatively little mathematics between 1912 and 1914. For the most part he had continued to be fascinated by Dirichlet's principle. There had been two more papers connected with it. One of these was especially significant because in it he stated the important lemma which had been implicit in his dissertation. In 1912 he and Born had collaborated on a paper concerning the result of Eötvös; and in 1913 he and Harald Bohr had written a paper together on the application of the theory of diophantine approximation to the zeta function of Riemann.

These mathematical works were not nearly so impressive in quality or quantity as those of some of Courant's contemporaries, most notably Weyl, who was already a professor in Zurich. However, Courant was considered for a position as an associate professor at Frankfurt; and his name appeared on the list of three sent by the Frankfurt faculty to the ministry. He did not receive the appointment (which went to Hellinger), but just having one's name on such a list was a first step up the academic ladder. In the summer of 1914, in spite of the increasing number of military maneuvers in which he had to participate as a member of the reserve, he had reason to be optimistic about his academic future.

All this changed on July 30, 1914.

At four o'clock on the afternoon of that day, Edith Stein sat in her room reading Schopenhauer's *Die Welt als Wille and Vorstellung* in preparation for a lecture by Adolf Reinach, a popular and gifted member of the Husserl circle. (In his own room Reinach studied the atlas.) A knock on the door announced a friend, who told Edith that a declaration of war was momentarily expected and that all lectures were being suspended until further notice. Almost immediately there was a second knock. It was Nelly Courant with the news that Richard had received his orders. She was leaving that night for Breslau, where she would wait with her father for the end of the war. Everyone, including Richard, was sure it would be over soon.

It was by now five o'clock, the hour for which Reinach's lecture had been scheduled. Edith closed *Die Welt als Wille und Vorstellung*, "which I never opened again," and began to pack. At seven-thirty she was at the Courant apartment. The train was to leave at eight. The carriage was already in front of the door. Nelly and Richard, who was going part of the way with them, were saying goodbye to each other in his study. "And they were not too quick," Edith reported. She was not yet aware that her cousin and his rigid, highly organized wife were finding life together increasingly difficult. Still, it was astonishing to her that Nelly would go back to Breslau before Richard had left for the war.

"In her place I would not have done so. But it was probably due to her concern about her father. And then also she was not in general like other people."

After accompanying his wife and his cousin as far as Kassel and placing them on the Breslau train, Courant returned to Göttingen to await mobilization. The next day the Kaiser's proclamation of war was posted on every street corner. Parents sent their children to look at it. Boys not out of the gymnasium stood in line to enlist. They had but one fear: that before they could see action, the war would be over. Although Courant was twenty-six, he was as eager to go as any boy—patriotically convinced of the right of Germany's cause and confident that the Fatherland would be victorious "even over a world of enemies."

War fever raged in the university town as virulently as it raged everywhere in Germany. Winthrop Bell, a Canadian student of philosophy whom Courant considered "the nicest person in Göttingen," was incarcerated as an enemy alien in the university "jail" on the third floor of the auditorium building. In the face of Klein's disapproval, his youngest and favorite daughter, Elisabeth, popularly known as "Putti," married Robert Staiger, a doctor of musicology and the conductor of the academic orchestra, just as he was about to leave for military service. The two Runge sons, neither of whom had yet completed the gymnasium, rushed to enlist and waited impatiently for the moment when they would don their field-gray uniforms. The two sons of the philosopher Husserl also enlisted.

There were, of course, some in Göttingen who were not infected by the fever. Hilbert saw no reason to swerve from his program of straightening out the physicists, and he continued to plan with Debye what was to become the famous seminar on the structure of matter. Runge, as the rector of the university, objected strenuously to the treatment accorded Winthrop Bell and insisted on transferring him to his own home, announcing that he would be personally responsible for him.

It was at this time, during the first days of war, that a neighbor woman was surprised to see Runge pacing up and down in her sheltered garden, obviously much upset. He told her that he could accept the fact that his sons, who were still only boys, should be swept up into the army; but it seemed to him wickedly foolish that the lives of young men of proved ability, like Richard Courant, should be risked.

To Courant himself the day of his mobilization and the week that followed were like "a marvelous dream."

"I shall never forget," he wrote in a diary which he shortly began to keep, "how many great and beautiful qualities appeared in people at this time, how everybody was especially kind to me, how pleased I was to find acquaintances—particularly Staiger—among my new comrades, how we were outfitted, how we left Göttingen on Saturday, August 8 (in circumstances dramatic for me), how we traveled endlessly on trains, how we marched in the full moonlight. . . ."

The diary was among the papers which Courant brought with him to America. It is contained in a lined notebook with a cardboard cover. The first entry is dated August 13, 1914.

After a long hard march, the regiment was spending a day near the Belgian border, resting, eating, swimming, sunbathing, and conversing. Courant, who was a non-commissioned officer with a rank equivalent to that of a staff sergeant, and his friend Walter Lohse, a tall dark-haired young man who played the violin in the academic orchestra, decided to shoot a few rounds from their pistols to remind themselves of "warlike activity."

Neither Adolf Reinach, nor Putti Klein's husband, nor the youngest sons of the Runges and the Husserls, nor Walter Lohse would survive the war; but Courant's diary begins:

"It seems to me that I am on a beautiful summer vacation."

SIX

AFTER AUGUST 14, Courant and his regiment were in Belgium, moving down the valley of the Meuse with the Second Army. Everywhere they saw dreadful devastation and found as a rule an intimidated population, willing and often eager to please them. But there were also instances of "stabs in the back" by the Belgians. Courant, noting these, added, "There are probably things on our side, too."

As he and his comrades proceeded farther into Belgium, sometimes quartered in villages from which all the inhabitants had been evacuated for fear that they might give away the German position, he found himself shocked by the destructiveness of his fellow soldiers. "Yet they are otherwise good people who are now playing the harmonium [they were quartered then in a deserted school] and singing chorales."

All day and all night he heard the sound of distant artillery fire. He wrote again and again, "If only I could be there!"

Quite soon, however, he and his company were detailed to another regiment and given the job of protecting a railway station at the rear.

"We must lie around this place and fill our stomachs while others bleed," he wrote disgustedly. "Perhaps our 91st Regiment has already been in battle!"

In the coming weeks, from time to time, his company was in positions where he expected to be called up for attack at any moment; but except for one rather unusual engagement, "in which not a shot was fired or a man lost," he saw no action. There was still more railway duty. At one point, when he was quartered in a customs house, he found a piano and began to play. He took innumerable photographs and developed them at leisure. Sometimes he found himself wondering if things were going as well as had been hoped. On September 20, 1914, he wrote: "Reports of victories which are passed on every few days remain without confirmation, but I have *every confidence* in our cause. It must succeed in the end. But I am afraid it will take much longer than anticipated."

News regarding promotions began to come through. Lohse's name had been proposed, but not Courant's. "The main thing against me is that I am considered too restless. . . . I must console myself with the knowledge that I have done my duty to the best of my ability."

50

In Maubeuge, which his company entered shortly after its surrender, he was shocked to see that many of the officers were shipping carloads of booty back to Germany. Some, selling cattle and coal to the inhabitants of the town, were distributing the money, "which belongs to the German State," among their men. His own company commander—"Mr. von H."— had sent for his wife. "[That] is humanly understandable, but should not occur. When she spoke to me, I felt real discomfort at her illegal presence."

Finally, on October 12, 1914, the order came for Courant's company to rejoin its regiment. The first night on the march they were quartered near Reims. Everything seemed especially beautiful to Courant. The landscape, the autumnal mood, the pretty little village. He and Lohse took a walk in the evening, and Lohse said thoughtfully, "One would really like to stay alive." Courant, recording the remark later in his diary, added in French, "We shall see."

The regiment had suffered heavy losses. They were alternately in the trenches and in the cellars of the village in front of Fort Nogent, which was under constant artillery fire. Courant found himself embarrassed by the contrast between what the regiment had done and the activity of his company in protecting railway stations, guarding prisoners, and collecting weapons.

At the reunion there was no action, however. He was reminded of children playing Indians. He was almost cozy as he sat in the dugout at a table before a fire and wrote in his diary, "I would be quite content if Mr. von H. did not systematically make life and work difficult for me." He felt constantly "pestered" by the commanding officer. "Unfortunately I am very easily disturbed in my mental peace by such antipathies," he wrote at another time. "The people with whom I live and work must be in harmony with me. Otherwise I am most uncomfortable."

In a few days, much to his surprise, his youngest brother, Ernst, suddenly appeared at the front. He had been working in Rumania but had rushed back to Germany to enlist. By chance he had been assigned to the same regiment.

Now that Courant was at last in the trenches, expecting action momentarily, he hoped that fate would give him an opportunity to distinguish himself. Instead, on November 1, 1914, less than two weeks after rejoining his regiment, he was taken to the nearby village with a high temperature. Two days later Lohse was brought from the trenches with similar symptoms. Both were transported to the field hospital, where it was determined that they were suffering from typhoid fever. Courant was delirious most of the time during the next three weeks; and Lohse, who was much less ill, stayed constantly with him. "Who knows whether I would still be alive without him?"

During his convalescence Courant "burned to get back to the front." There were two reasons for this desire, he confessed to himself: first, he found sitting in the rear "unbearable" and, second, "because of a little ambition on my part."

Finally, moustached and bearded, he was permitted to return to active duty on February 5, 1915. He felt that he came back "more serious, but gladly and eagerly.... Who knows what is still ahead?"

In a few days he and the regiment were on their way to the Argonne Forest.

While Courant's battalion was setting up its position to a constant "hellish concert," his brother Ernst's battalion lost more than 220 men in a single day. Courant went to look for his brother and to his relief found him well and unharmed.

"His company chief told me that he had deported himself 'quite heroically'. . . . I was very proud of the dear modest youngster, who did not wish to talk about his achievement."

Everyone seemed to be sick, at least with a cold.

"But in spite of my bad cold and torn boots, I am quite well. Also my nerves have suffered less than those of many others."

A week later, after spending two days and nights building a new position under heavy fire, without sleep and without cover, he was running a fever and not able to speak. Since Lohse had arrived to take his place, he allowed himself to be sent to the rear.

"What a strange feeling it was when I left the forest by field train and saw again houses and other people!"

He was in the hospital this time for ten days and had leisure to bring his diary up to date: "In the Argonne there is no soldier that is not fed up with the war. . . . But that doesn't mean that we are disheartened. . . . No German soldier would allow the thought of defeat to enter his heart."

A few days after he returned to his company, orders came to move on to Vauquois.

Vauquois was a beautiful little village on a promontory to the east of the Argonne Forest. It had been in the possession of the Germans for some time, but the French had made a series of violent attempts to regain it. Both sides currently shared the ridge. Losses were running to more than fifty per cent.

Courant had not forgotten how, leaving the forest for the hospital, he had looked up at Vauquois—"an ocean of smoke, lightning . . . the most violent artillery fire concentrated on a small height"—and had thought, "Thank God, I am not up there." Now his regiment waited below Vauquois for the order to attack.

"I am dulled and stupefied, look neither ahead or behind, do nothing except take a few photographs. . . . What seemed interesting and important and unusual to me only a few months ago is of absolutely no consequence now. The word 'culture' sounds almost like a sneer."

When Courant and his comrades arrived at Vauquois, they were deployed on the slope and given a few moments to inspect the situation. The demarcation line between the Germans and the French ran along the highest point of the ridge, where the church stood.

Three weeks later—after almost constant fighting—Courant described in his diary what he had seen in those first moments:

"The interior of the church was filled with rubble and bodies and parts of bodies, one corpse naked, burned by a flame-thrower. At the right of the church, the position curved around the famous linden tree toward the back trench, which . . . utilized the wall of the church cemetery as its bulwark. The connecting saphead led across two coffins. In the trench there were several corpses. The heels of one of ours were sticking out, the heads of others. . . . The most dreadful sight was an arm in field grey with a white hand which protruded from the right wall of the right saphead. Everywhere we dug, we dug into corpses. More of ours than of the French. Whenever a mine exploded, pieces of flesh fell into our position. Shreds of a dead Frenchman hung from the branches of a tree. . . ."

Courant did not again take up the diary which he had started at the beginning of the war. Instead, in August 1917, he completed the record of the rest of his active duty in summary form. The following details are taken from that account.

After almost three weeks' rest in the Alsace, the regiment was called up again and the company installed in an advanced position, surrounded on three sides by the English, west of La Bassée. They had no connection with the rear, since telephone lines were immediately put out of commission by artillery fire. During six days of fighting they lost more than half their men.

Describing these days—"None of us believed we would ever emerge from this hell!"—Courant remembered a night which he had shared with Lohse. Bayonetted rifles in hand, they had crept back and forth, stumbling over the dead and wounded, trying to encourage "and console" their men. In the course of this activity they had collided. Across the grisly scene they had smiled at each other "like augurers" and "without saying a word about the situation we were in, we sat for a while there in the trench and talked about Göttingen . . . until a bursting grenade dislodged a stone which knocked my helmet from my head and my gun from my hand."

It was during this period of active fighting that one of Courant's ideas for relieving the technical difficulties of trench warfare began to be put into operation. Some months earlier in Laon, looking at the cathedral against the horizon, he had had the thought that mirrors could be used to enable the men in the trenches to see what was going on above ground without exposing themselves to enemy fire. In La Bassée, as a result of his suggestion, trench mirrors in great quantities were being produced from requisitioned mirror glass.

Trench warfare had also made him realize how valuable it would be to have a means of communication that would not be vulnerable to artillery fire and could be operated under full cover. Telephone lines, he knew from experience, were the first casualties. The light signal system, which was also sometimes used, required a straight-line connection and also required that a man leave his cover. He began to think about resonance synchronization with electromagnetic signals which would utilize the earth as a conduit.

He talked about his idea to Lohse and his fellow officers. (He had finally been made a lieutenant after the hard fighting of May and June 1915.) Later he spoke to his battalion commander and found him also receptive. He was surprised how unusual even his small amount of technical knowledge seemed to be in the army. He suggested that he should be sent to Göttingen to consult with the scientists there about the feasibility of earth telegraphy, and he shortly received a leave.

An overnight journey took him to Hannover, where he would catch the train for Göttingen. Then he would see Nelly, who had returned to their apartment. He went for a walk to prepare himself "a little bit" for what awaited him and almost missed his train.

In Göttingen, he was relieved to find Nelly not in the apartment. Leaving his things, he went to the Runges' home. They had lost their youngest son in the first months of war, and they greeted the returning Courant like a son. As Courant had expected, Runge was able to make many helpful suggestions.

Runge's daughter Nina went to look for Nelly.

"They both returned together [and] I received a rather frosty greeting. The scene ended when Nelly and I took a carriage and returned to our apartment, where she had invited three maiden ladies to live with her during my absence."

He felt that a break with Nelly was imminent, but he nevertheless enjoyed his leave. After discussing his ideas for earth telegraphy with Runge, he went to Berlin to inquire at Telefunken, the German telephone company, about the possibility of using small wireless stations at the front. Count Arco, the head of the firm, was of the opinion that the necessary distances could be achieved only by antennas three to six feet high. He showed Courant some experimental models for earth telegraphy but pointed out that they could achieve a distance of 500 meters only with a great expenditure of energy.

With one day of leave remaining, Courant returned to Göttingen, very discouraged. To his delight he found that in his absence Runge had brought Peter Debye and his assistant, Paul Scherrer, into the project. The next day the four men worked together. By evening they had succeeded in sending a message from the physics institute to the Leine River, a distance of some 1500 meters. They had as yet no usable result, but they felt they had a sure expectation of a satisfactory solution to the problem. Courant returned to the front with a statement to that effect signed by Runge under the official seal of the university. He asked immediately for another leave so that he could carry on further experiments.

When he returned to Göttingen in August 1915, he learned from Debye—the first person he met on the street—that during his absence successful experiments had been performed with the pendulum transformer of the German telephone works. For the next few weeks he worked diligently with Runge, Debye, and Scherrer in the fields around Göttingen.

Nelly had left for Berlin, and he spent most of his free time at the Runge house, playing the piano with Nina Runge, who had studied the violin in Hamburg for the past few years.

"I was just amazed at the chutzpah with which he tackled anything that was attractive to him," Nina Runge, now Courant, told me. "You know what chutzpah is? Nerve! He never suffered from any inhibitions because something was too difficult for him, but tried everything. He somehow succeeded in making a piece intelligible, getting the spirit of it, no matter how imperfect his playing was. On the other hand, really practicing, finding a method of improving something that was too difficult for him—that he couldn't do. But I learned something from him. I made up my mind that pieces imperfectly played are not spoiled by this fact as long as one understands them. I learned not to be afraid of them. Why think of the listener? We played for ourselves alone!"

On the last day of Courant's leave, Scherrer presented him with a box he had built to hold the earth telegraphy apparatus, the result of three weeks' work. It could transmit messages over two kilometers. It seemed to Courant "a magic box," which would produce good prospects for his future military career.

After his return to his unit, he drafted a letter about the instrument to the General Command and, on his own, trained a few men to use it. He hoped to be able to arrange to have Runge and Debye come to the front to conduct further experiments. But days passed without a response to his letter. Then one morning while Courant was demonstrating his apparatus for some friends, General von Etzel, the brigade commander, passed by. He was so excited by Courant's idea that he placed him immediately in contact with the communications division. That day Courant wrote Runge an exuberant letter.

"Concerning the organization, I have a daring plan . . . ," he confided to Runge. "I want to set up a little factory behind the front to produce such magic boxes. This is not actually as crazy an idea as it may at first appear to be. First, we can easily obtain workmen here, mechanics, carpenters, electricians, materials, tools. Third [sic], we can avoid all bureaucratic difficulties. Fourth, we will not have to depend upon the goodwill of industry. Fifth, it will be cheaper. Sixth, the other factories will have time to do something else. The question is only whether *I* can manage to establish such a factory. . . . The military still seems to be less impressed by the earth telegraphy apparatus than by the fact that a young lieutenant can get a *Geheimrat* and a *Professor* to interest themselves in something which this lieutenant considers important!"

While Courant waited impatiently for his transfer to the communications division to arrive, artillery fire became more and more intense. It was clear that during the anticipated offensive his regiment was going to find itself in a "particularly uncomfortable" position.

He finally received his transfer on the same day that his regiment received orders to march. There was no question where his responsibility lay. All dreams of developing the earth telegraph seemed to have been wiped out; but he decided to take his magic box to the front, just in case it could be used.

The regiment marched to Douai, where it was held in readiness for the attack. Courant heard that Ernst had been taken to the hospital with what was thought to be appendicitis. He hastened to see him, found him not so ill as he had feared, and rushed back. In the hour he had been gone, the regiment had received its orders, each battalion marching off to a different railway station.

Courant took his magic box and went as fast as he could to the nearest station. The train left in the dark in a thunderstorm. When the troops disembarked, his battalion was not to be found. Taking up his pack, he set out in a soft rain. After awhile the rain stopped and the moon came out. An ambulance picked him up and took him to Sallaumines, where his battalion shortly appeared.

All night he stood watch. He had the impression that in the section where he was the fighting was over. But suddenly there came a call to march. Soon he and his comrades arrived in a region of utter confusion. They were given ammunition and told to leave behind any dispensable baggage, which must have included the magic box.

As they proceeded, they met almost no one except, in a dugout, a few men who told them that the English were quite close. It was so dark that they could see only a little distance ahead. Sudden machine-gun fire announced that they had made contact with the English front line. There was a brief exchange, but they were able to take cover without much loss.

In the daylight Courant made a tour of inspection. Corpses were everywhere, in some places so close together that he had to step on them to get through. In front of the position there were moaning English wounded who could not extricate themselves from the barbed-wire entanglements. Toward evening he was able to have some of them pulled out and sent to the rear. Among these were two Scotchmen, with whom he "became quite friendly."

He had had no sleep for thirty-six hours, and he curled up in a hole and slept, as deeply—he later recalled—as he had ever slept in his life. Early the following morning he was awakened from "the pleasantest dreams" by Lohse, who told him the battalion was to attack Fosse 8.

The advance proceeded at a fast clip across a wide plain; but there were still losses, and screams to right and left from the injured.

All during the advance Courant had the distinct feeling that something was going to happen to him that morning. It seemed to him that he almost waited for a bullet. He found to his surprise that this premonition, instead of deterring him, made him more calm. Now one more advance would bring him and Lohse and the few other men still with them to the comparative safety of the Zeppelin Dugout. Courant, lying somewhat in front of the actual line of his men, prepared to give the command. Suddenly, out of the early morning mist and smoke, a line of English infantry appeared. He jumped up to rejoin his men. Almost instantly he felt a violent blow to his right side and fell to his knees. He was sure he had been shot in the stomach and would be dead in a short time.

". . . following some instinct of survival, I tried after a few minutes to crawl toward the rear," he wrote two years later, recalling the event. "I could not do it by myself, and two men from the neighboring battalion came to my aid. One of them was shortly hit in the leg, and immediately afterwards I was hit by a second rifle shot, which took great pieces of flesh from my left lower arm. I was beginning to lose consciousness because of loss of blood, although my other companion tightened a tourniquet about my arm in a shell crater. The man who had been hit had his leg dressed while the other man rolled me into a coat, put me on my back, and dragged me along the ground with a piece of cord from a bread sack. In my dazed state I tried to help by making rowing movements in the air with my hands and feet. We advanced only by inches. The noise, bullets and shrapnel, was infernal. The English advance was imminent. I remember that during the hours of being pulled I was constantly concerned with whether we were actually going in the right direction, since no definite line of demarcation separated us from the English. I was overjoyed when

I finally saw the familiar pile of straw in front of our barricade, and I
I was taken by some strong arms and placed on a stretcher and carried
across the trench to the dugout."

In one of those eerie coincidences which sometimes occur, I was translating the preceding paragraph of Courant's wartime recollections when I
received a telephone call that Richard Courant had died in New Rochelle on
January 27, 1972, at the age of eighty-four.

SEVEN

THE BULLET on September 27, 1915, put an end to Courant's career as an infantryman; but it had not penetrated his stomach, as had at first been feared. He spent the month of October in the military hospital at Essen. During this time Nelly came to see him and told him that she wanted a divorce.

At the beginning of November he received a convalescent leave. Nelly had already let the apartment in Göttingen, but the Runges invited him to live with them. He had the strong feeling "that this stay in Göttingen was going to be a test to see if I was worthy of having my life spared for the second time in this war."

A month and a half later, when his leave was almost up, he had to admit that he had failed the test. He had sought out company—"like a drinker seeking out alcohol." He had wasted time—"There was rarely an evening when I did not have to say 'a lost day.'" He had become involved with a young woman, a student of mathematics and physics—"I have plunged myself even deeper into guilt." On December 16, 1915, the day after she left Göttingen, he began still another journal for self-examination and self-improvement. Although the refrain of "wasting time" and "being always with people" is the same as in his earlier journals, there is a much greater urgency. He seemed to himself "a bankrupt" in life. "If I don't put an end to this carelessness and weakness, it will be the end of me."

During his convalescence he had begun to think again about his academic future. There had been a hint that he might be called to Bonn as a professor; but, going for a walk in the snow one evening with Landau, he learned that he had in fact no chance of getting the appointment. He did not feel that he deserved to be a professor "either through diligence or devotion to science," but it made him angry that people of his own age and gifts who were not soldiers were being preferred over him. When he returned to the Runges' house, there was a telegram informing him that he was to report for duty in Berlin at the first of the year. Nina Runge was still up, and he sat for a long time talking with her.

"I spoke a great deal about myself, so much in fact that I feel uncomfortable about it now. It is very difficult for me at this time to be really open with other people, especially face to face with people whose respect I want as much as I want Nina's. . . . Besides, it is perhaps pointless now, or even harmful, to aspire to close friendship with Nina. At the moment I am fundamentally unfit for human intercourse. . . . Perhaps people like Nina, through their way of life and their actual existence, can be of some encouragement to me now, but more than that I cannot expect from them."

At the beginning of 1916 he established himself in his parents' apartment in Berlin. Siegmund Courant was currently doing quite well with the sale of concentrates which were being sent to the front. Ernst was at home on leave. Courant found that the more he got to know his youngest brother, the more fond he became of him. But he continued to complain that he was *never alone*. He felt that he must have solitude to find himself and that the keeping of his journal should force him into that situation.

After a few entries, however, there is a gap of almost three months.

Outwardly he carried on. He established contact with people at high levels and had some success in convincing them of the value of his proposal in regard to earth telegraphy.

He and Nelly obtained a divorce. The final decree, dated February 16, 1916, stated that the marriage had been breaking up before the war began and that Courant was the party at fault. Nelly resumed her maiden name and became a teacher of mathematics in Essen. She never remarried. During the Second World War she died in the gas chambers of Auschwitz.

When Courant took up his journal again, he was in Heiligblasien to conduct experiments with his telegraphic device at the front. He no longer had the enthusiasm he had had in the past for the project. In many cases earth telegraphy was surpassed by high frequency. He had tried, so far without success, to get control of that too. Objectively, he felt, there was no doubt in his mind that it would be good if he were able to take the whole business in hand and thus counteract "the somewhat shady and uncooperative maneuvers" of business firms and ambitious officials behind the front.

"But I must confess to myself that I have still another motive in my heart," he wrote in March 1916. "I would like to survive uninjured and *on this account* to be active in other ways than in the trenches and in assaults as a platoon commander. That is perhaps too crassly expressed, and it might be different if I knew that on my return to the regiment I would have a relatively congenial situation in relation to my comrades. But what I said is so—I mustn't lie to myself."

The fact was, he conceded, that he would seize "with both hands" any opportunity—"which in my opinion is not unworthy or contrary to duty"—in order not to endanger himself beyond "acceptable" measure. He felt that his conscience would be clear if the earth telegraph project would just begin to move.

"I have no martial spirit, but I do have a very distinct social consciousness and on this account despise every sort of shirking. This is what made me a very passable officer—to share the responsibility for the actions and destinies of others, not to leave others in the lurch nor to expect more of them than I would of myself."

The next day's entry is in a completely different spirit. He had received orders to leave for Stenay in the Ardennes.

"And who knows," he wrote, "if I shall not soon have the opportunity to test earth telegraphy in action at Verdun . . . ? This turn of of events has electrified me. All the silly reflections are thrown out. Let's get on with it. That is the right way."

At the front, in the course of making some tests to obtain an idea of sounds coming through the earth against which the telegraphic signals would have to compete, Courant was startled to hear signals similar to his coming from another source. Since there was relatively little action at the time, a patrol was organized which, under the protection of "a little artillery fire," crossed over no-man's-land to the unoccupied front position of the French. Jumping down into the dugout and looking around, Courant discovered exactly what he had expected—an instrument similar to the one he had developed, but superior in some ways, and an instruction book for its use.

After this discovery there was greatly increased interest on the part of his superiors in Courant's apparatus. In his journal he begins to show concern with the effect of interpersonal relations on the project and with his own position in relation to it.

He was attached to the *Funkerkommando*, or wireless command; and as an infantryman (which he still considered himself to be), he found that he had a quite different attitude toward tactical questions from that of "the gentlemen of the *Funkerei*." He knew what it was like in the trenches. With approval from general headquarters for the experimental installation of some devices at the front, he had another concern. Now he must see that he was not forced out by the *Funkerei*, "for that would not be good for the project." He found it "idiotic" that the project should be tied up with the *Funkerei* at all. The captain was a man of "staggering narrow-mindedness." "A big blow-up" was coming. "But I shall not let the project suffer because of it."

Two weeks later what he had feared actually took place. The captain tried to force him out and replace him with an officer of his own.

"I was very upset—up to the borders of the militarily permissible in expression of my anger—but then, in the interest of the really important project, I swallowed the insult and went along to Haraumont. [Here the heaviest fighting at Verdun was taking place.] I oversaw the installation of the apparatus and spent several days in great danger."

By the middle of May favorable reports on earth telegraphy began to come in from Verdun. Plans were made to manufacture a large number of the devices. Courses were to be given so that officers would be trained in the techniques required. The chief of all field telegraphy was coming for an inspection.

Courant felt that at this point he could hardly be pushed out.

"On the other hand, recognition will essentially be steered to others. Everywhere people are now ready to take credit for the project themselves. . . . That this so easily happens is partly the result of my politics in launching the matter—namely, I have suggested to people of rank or influence that it go out as if from them. Thus I have been able to mobilize many forces where I myself would have been powerless."

The inspection, a few days later, came off successfully. The entire front seemed to be demanding the apparatus. The chief of telegraphy sent out a circular to all staff officers in which he singled out Courant for praise. There was the possibility of an Iron Cross, *First Class*.

"I must confess that now in this moment when everything appears to be secure my personal vanity and ambition awaken."

In spite of the successful outer circumstances of Courant's life, the inner situation had not improved.

"Often I observe in myself an almost pathological indecision, especially in the small things of daily life. I can, for example, vacillate over whether I should go to the theatre until it is finally too late to go and the evening is lost anyway. Most things concern me so little . . . that there is not much difference between *yes* and *no*."

In his journal he continued to castigate himself.

"[Nelly] was completely right when she said once that at the bottom of all my difficulties is the fact that I have no true relationship to my subject—and that is because I too much pursue every aspect of success rather than the subject itself."

He felt that he had been a completely different, purer, much younger person three years ago.

"I can hardly remember that person, but I shudder when I think of the difference."

In July 1916 he received news that his brother Ernst had been killed in close combat in the fighting around Verdun. During the latter part of the year, in the course of his extensive travel for the purpose of supervising the installation of earth telegraphy devices at various parts of the front, he stopped to visit Ernst's grave in Ablain as often as he could.

"But," he reported nine months later when he finally took up his journal again, "it has not helped me."

In contrast to the inner agony, the outer life which Courant led at the beginning of 1917 was one of unheard-of freedom for a young lieutenant.

"For about three weeks I have been so intensely occupied with military matters that I have hardly had a moment to myself—just as you would expect of an officer of the General Staff," he reported to Hilbert. "Of course I cannot tell you more about these matters. Only this much. I was at the Somme until January 10, where I saw a great deal. Then I was called to Headquarters in Silesia to give reports and make suggestions, with some considerable success. From here I went as the personal emissary of Headquarters, via Berlin, to the Western Front, where I am waiting for a telegram from an important personage mentioned daily in the press."

He reveled in the importance he had and, even a half century later, described with pride how at Siemens, a firm which is sometimes called the General Electric of Germany, he was taken to the dining room of the highest echelon, where the directors of the company entertained their guests. He discovered that Crown Prince Rupprecht of Bavaria, to whose division he was attached, was "very nice"—that Count Arco, the head of Telefunken, was "a very intense and very imaginative and active person" —that Otto Arendt, the director of the postal service, was "a really very unusual type, a very good engineer and a very intelligent man." He also learned that as an independent scientist he was able to talk to these people at the top "as an equal."

When he stopped in Berlin on his way to the front again, he heard that there was a possibility that Hilbert might leave Göttingen and come to the capital.

"You can imagine how closely the news touched me," he wrote to his former professor. "My first thought was—if you accept, it will be the beginning of the end of the Göttingen era, but possibly also the beginning of a new era in the life of the mathematical and physical 'world'. . . . But it seemed to me that it would be better for you personally—and for this reason better for science also—if you were to remain in Göttingen, to which you are bound by so many ties. . . . Although I have not been a regular member of this circle for quite a long time, my close ties and great attachment to Göttingen really depend upon your presence there. Your departure would leave a painful gap for me. Many others will feel likewise."

A week after Courant wrote this letter to Hilbert, Germany began to carry out its previously threatened policy of unrestricted submarine warfare.

At the beginning of February 1917, Courant received orders to go to the front and establish facilities for training a number of men in the use of the earth telegraphy apparatus.

He had no idea how to set up such a project within the framework of the military, but he put himself in the hands of a sergeant who seemed to know everything. The sergeant gave him a few samples of orders and let him use a typewriter to compose something similar. Arriving at the last point in the order, Courant had the idea that it would be useful for him to have "an assistant." He put down the name of Wilhelm Runge, who was currently at the front and whose military address he knew from Göttingen.

At this time young Runge was not someone whom most people would have selected. He had almost nothing to recommend him other than the fact that he was Nina's brother and Professor and Mrs. Runge's only surviving son. Bernhard Runge, killed in action in Belgium at seventeen, had been precociously gifted. The family could only hope that Wilhelm would turn out to be "a late bloomer." He had been unsuccessful in the gymnasium in Göttingen; and his father ("who was very patient with me") had sent him to a school in another town, where he had done only a little better. He had been in the trenches for two years and, although exposed to officers' training, had failed to make the grade.

Courant had an intuitive feeling that Wilhelm had potential. At the same time he saw an opportunity to repay the Runges for their kindness to him by removing their son from almost certain death.

The commanding officer objected to the last line of Courant's order—it was highly irregular, he said, to request a specific person, and particularly one from another division. But Courant insisted that he needed someone he knew and had confidence in. Sergeant Runge was "a very gifted young engineer."

Wilhelm Runge later became chief director of research at Telefunken. During the summer of 1971 I talked to him in Ulm, where, although retired, he continued to oversee Telefunken's institute for basic research. At that time he described to me how in the early spring of 1917 he helped Courant establish facilities for training men in the use of earth telegraphy at Cambrai in northern France. He also told me something of their life in the occupied city.

"We had an apartment, Courant and I, in a private house. There was a grand piano there, a very beautiful thing, and Courant found some music for quartets and trios. There was one piece I liked very much. [It was Beethoven's "Serenade for Flute, Violin and Viola, Opus 25," Nina Courant told me immediately when she heard her brother hum the opening bars on my recording of our conversation.] Afterwards I always used to use it as a whistle call when I visited Courant or when we passed on the street. In my memory of that time, I always see Courant sitting at this grand piano and bringing some piece into the form that you can play on the piano." Runge apologized for talking so long. "But I wanted to bring you this picture of Courant, in this occupied land, playing the piano."

When Courant received orders to go to other parts of the front, he left the installation and instruction connected with earth telegraphy in charge of young Runge.

"Once again I am on my way," he wrote to Hilbert on March 4, 1917. In spite of the success of his efforts, there were things about his present activity which were nerve-racking. "For the simplest, most obvious things it is a constant battle against stupidity and ignorance. In most cases it is I, a low-ranking young lieutenant, opposing all sorts of high officers." There was "a mountain of work and such enormous responsibility as cannot even be imagined by someone in civilian life." He could barely remember what it was like to have had enough sleep. "Before long," he concluded, "things will probably be even worse."

Yet it was at this time that Courant apparently began to think about mathematics again. In the letter to Hilbert he mentions that he is sending along a manuscript. It is not specifically identified, however.

"I am increasingly convinced that in business and in the academic field those who give their health and strength as soldiers are being short-changed while others at home are advancing their careers without much effort."

Several weeks later he once again took up his journal, this time after a lapse of ten months. His project of earth telegraphy had had much greater success than he could ever have dreamed. Millions had been spent on it ("although, to be sure, it is almost too late"). For the past months he had been traveling in the field under orders from the chief of telegraphy "with rather great authority." Yet he found little satisfaction in his triumph.

"What do I care? Inwardly I have not been able to pull myself together."

Recently, however, he confided to his journal, he had begun to think that there might be a way out for him—a wife.

"But of all those I know it could be only Nina. Yet how could I dare now seriously think of binding Nina to me? I would then have to be so frank with her that she could not preserve the necessary respect for me. In spite of that, I have still thought that at the next opportunity I would at least ask her if later there would be a chance of her becoming my wife."

Now to his surprise a letter came to him from Nina.

While he had been thinking of her as a potential wife, she had been thinking of him as a potential husband. She knew he had been in love with the other girl in Göttingen, but she thought that he was over that. Of all the young men she knew—she told me—he was the only one to whom she could imagine herself as married. She sensed the turmoil within him.

In a brief, straightforward letter she wrote that if he felt she could help him, she was ready to become his wife.

Courant, receiving her letter on March 23, 1917, made what was to be the next to the last entry in his journal:

"In spite of doubts and difficulties, all is now basically determined. For myself I know that everything depends on whether I can succeed in grasping this hand that has been stretched out to me from a better and purer world."

A few days later, just when the deep personal dissatisfactions which had tormented him for so long seemed on the verge of being resolved, he experienced a significant setback in his military career.

Very early in his activity he had gained access to the chief of telegraphy and won his personal support. In the introduction of earth telegraphy at the front, however, he had felt constantly hindered by the jealous opposition of the entrenched staff of the regular army. The struggle is reflected in over a hundred official orders covering the period between January 1916 and March 1917.

At the end of March he suddenly found himself replaced on an important mission by another officer. He sputtered angrily in his journal:

"This man—without any knowledge of the project, without the slightest general or special understanding of it—is being sent only because he is a captain in the regular army and I am only a little lieutenant in the reserves. And this happens just at a time when it will spoil everything! My position in relation to the chief of telegraphy has evidently been undermined. I have too frankly expressed myself. Probably I will soon be completely pushed out."

The entry, dated March 26, 1917, is the final one in the journal for self-examination which he had begun during his convalescence in Göttingen. Eleven days later the United States declared war on Germany.

During that summer of 1917, Courant's old comrade in arms, Walter Lohse, was killed in action.

EIGHT

URING the remainder of the war Courant no longer played an active role in the earth telegraphy project. After a period in Berlin, he was sent to Ilsenburg, a little town in the Harz Mountains, not far from Göttingen, where a school for training men in the use of the apparatus had been established.

The new situation was not without advantages. He could make frequent trips to see Nina, to whom he was now engaged; and he had enough time to write her long letters, often two or three a day.

Hundreds of men attended the school—"some of them quite intelligent, but most of them not." It occurred to him that a booklet of written instruction in regard to earth telegraphy might be helpful—"some little abbreviated exposition of what it was all about and how it should be used"—and he proceeded to produce a manual. It was his first attempt at expository writing and revived earlier ideas about the need for improving mathematical books.

In a letter to Nina in September 1917 he described how, having in mind "a publishing project," he had approached Arnold Berliner, the editor of *Naturwissenschaften,* for an introduction to that journal's publisher, Ferdinand Springer. On September 28, 1917, he wrote to Nina, "Today I must report to you before everything else that I have finally spoken with Springer and that everything goes well—still, of course, in the preliminary stages but completely in accord with my ideas."

Two days later he added, "The thing with Springer is something! I am afraid only that it is going to rest quite heavily on my shoulders. It will have to be very well thought out before it is made public. Therefore, please, the strictest silence! With everyone!"

Ferdinand Springer, who with his cousin Julius directed the Springer firm, had studied at Oxford. He had planned originally to become a diplomat, but he had later become convinced that an independent and free man could achieve much more than one who labored for someone else, even a great nation. He and his cousin now divided responsibilities in the family publishing business. Julius handled the engineering sciences; Ferdinand, medicine and the natural sciences as well as some areas of the arts.

Springer and Courant shared a common energy and optimism; and, as an additional bond, both had seen action in the war, Springer having served as an officer in the artillery until he was severely wounded in the foot. They took to each other immediately.

The publishing project which Courant had in mind was a series of up-to-date mathematical monographs in fields of mathematics which were particularly relevant to physics. They should make it easier for the physicist reader to comprehend mathematical ideas and methods by sparing him tiresome detours. They should also impart to the mathematician reader— along with the mathematical theorems—an awareness of the connections between mathematics and its applications.

Courant proposed to act as chief editor of the new series and, as such, to select or approve the topics and the authors. He also expected to do some writing himself. He had in mind a book which, he was certain, would have much to recommend it to Springer. He planned to approach Hilbert about the idea of a collaboration—a book based on the notes of Hilbert's lectures on partial differential equations and other areas of mathematics related to physics. As his share of the collaboration, Courant would do all the work of preparing the notes for publication. He was convinced of a need on the part of physicists for such a book; and then the combination— *Hilbert mit Courant*—that would be nice too.

Springer immediately saw in Courant a useful ally in entering a comparatively new field of scientific publishing. Although Courant was not yet professionally established, he had contacts with an impressive number of people in his subject.

Surprisingly, there was a large amount of important scientific work being done in Germany during the war. Courant insisted that it was vital to get this work into print as soon as possible. At the end of 1917, the Springer firm announced the establishment of a new journal, the *Mathematische Zeitschrift*. The first issue appeared in January 1918. The second issue contained a paper by Courant "written while in the army."

Courant had long been attracted by Rayleigh's *Theory of Sound*, which he considered full of original insights and suggestions; and in later years, when he was teaching again, he made frequent references to Rayleigh in his lectures and always urged his students to read the *Theory of Sound*. ("Which they rarely did," Friedrichs admitted. "It was very unsystematic. Nothing streamlined. Nothing put in a general framework. I couldn't read it. But Courant admired it very much.") In his 1918 paper he proved a theorem suggested by Rayleigh to the effect that among all homogeneous membranes having a given perimeter and a given tension, the one that is circular has the lowest fundamental tone. In the course of this work he realized that similar methods could be applied to the problem of the Lorentz conjecture, and he took up that problem in another wartime paper. His real work on it, however, was not to be done until after the war.

By the beginning of 1918, when Courant was returning to mathematics, the German high command knew that the war was lost; but the German people, ignorant of the real situation, were still thinking in terms of complete victory. In the spring of 1918, Ludendorff launched a desperate offensive, which he called "the Kaiser's battle." At first there were victories, but none decisive. Then—in July—the French struck back. It was at this time that Courant went to Göttingen to talk to Hilbert about the proposed book on the methods of mathematical physics.

"The whole concept so closely conformed to the secret of your own teaching success—at least to part of it," he explained later to Hilbert, "that I felt as your student I could summon up the courage to make such a proposition to you."

By October 1918, to facilitate an approach to the Allies with a request for an armistice, Germany had become a constitutional monarchy with Prince Max of Baden as imperial chancellor.

"Much could have been less painful and less trying if it had not been for the indolence and stupidity of the intellectuals and the frivolity of 'leaders' who permitted us to go to the very brink of the abyss," Courant concluded in a letter to Hilbert, devoted for the most part to their future collaboration. "Perhaps we will still have to face a rather painful purging up to the topmost levels. . . . Nevertheless, the path now opens again for all those things which make human life valuable. That this path will be taken and that hope will not fade is the task for all of us who have up to now been so critical. . . . I am looking forward to the moment when I can take off 'the king's coat' . . . and return to the Germany of Hilbert and Einstein. . . ."

Almost immediately the "painful purging" he had predicted began to take place. There was mutiny and revolt. Soldiers and Workers Councils—in imitation of the Russian soviets—were organized all over Germany. Courant himself was elected to head the Ilsenburg council. It was extremely unusual for an officer to be chosen by his men; and there were some people who were inclined to attribute Courant's selection more to left-wing sympathies than to personal popularity.

The creation of such groups as the Soldiers and Workers Councils didn't contribute much, in Courant's opinion, to clarifying the situation. Two weeks after the armistice of November 11, he described to Hilbert how at inspection nobody was sure who should give commands—the Council after consulting with the Commander, or the Commander after consulting with the Council.

"The majority of soldiers do not show any particular enthusiasm for the new order of things. They take it as an inevitable fact of life, just as they did the old situation in relation to authority. All they want is to get home as quickly as possible and to be free from further war or disturbance. This means that the moderating influence which the soldiers would have had on political developments is shrinking, and the more active and radical minority can easily gain the upper hand."

Courant went to Berlin to try to straighten out the whole confused business of the demobilization of his men. The situation in the capital seemed to him like that in an airplane when the pilot has died in flight.

He met Einstein—"a really wonderfully noble and pure personality"—and promptly "put him in touch" with Kurt Hahn, who had become the personal secretary of Prince Max. "I believe that such people [as Einstein] can do more for us now than experienced professional diplomats," he wrote to Hilbert.

Then, in the midst of post-war revolution, he sat down with Ferdinand Springer on November 24, 1918, and signed the contract for the series of books to be known all over the mathematical world as the Yellow Series.

Courant received his discharge as a lieutenant in the reserves. Among his papers in New Rochelle there is a decoration for having been wounded in the service of the Kaiser. There is also a Distinguished Service Cross, Third Class. But no Iron Cross.

The Göttingen to which Courant returned in December 1918 was very different from the one he had left. The number of students in classes ranged from forty to eighty as compared with several hundred in pre-war days. There were few foreigners—a scattering of Swiss, Scandinavian, and Dutch—but a large number of women. The men were mostly wounded veterans. Four of these were blind. Lecture halls were insufficiently heated and lighted. Everybody—professors as well as students—was at the least a little hungry. But the university was far removed from the violence of the capital.

Hecke had succeeded Carathéodory in Klein's chair. Klein continued as the scientific leader of the university, although he was almost seventy and failing in health. Hilbert, however, was taking a more active part in administrative affairs than he had in the past.

There were some new younger people. Alexander Ostrowski was helping Klein with the editing of his collected works. Adolf Kratzer was now Hilbert's assistant for physics; Paul Bernays, his assistant for logic—Hilbert's new developing interest. Emmy Noether was delivering lectures in Hilbert's name and with his support because, since she was a woman, she had not been permitted to become a *Privatdozent*.

Courant had been awarded the title of "professor" by the university while he was still in the army. It was a purely honorary title, he explained to me—"a Red Eagle, Third Class," the lowest civilian decoration. The only academic position available to him upon his return was that of assistant to his prospective father-in-law.

Like other discharged veterans, Courant still wore his uniform but with the buttons that had borne the two-headed imperial eagle removed. To people who had known him earlier, his new assurance and knowledge-ableness were impressive. Even to the students, it was apparent that Klein and Hilbert were relying heavily upon him.

In the middle of December they dispatched him to Berlin as a member of a commission of faculty and students to take up the inflammatory question of increased student participation in the affairs of the universities. (One radical student organization had already proposed that professors be elected by the vote of the students.) Heinrich Behnke, who was in Göttingen at that time, recalls Courant amusing a group of young people with his vivid description of the adroitness required for such a journey to the capital. "But after ten days Courant was again in Göttingen, and we students had the impression that he had obtained the best possible results —the machinery of the university continued to function."

After the abdication of the Kaiser, none of the many political parties which existed was powerful enough to rule. All fought passionately to do so. Although as a boy Courant had entertained the idea of "becoming a revolutionary," he felt that he had seen enough of revolution. In future years he was to say that while there must be change, there must never be *discontinuities*. The best hope for Germany seemed to him to lie with the Social Democratic Party. It supported a social organization based on Marxism, but advocated winning power by taking over control of the bourgeois state rather than by overthrowing it.

Courant had long discussions about the future of Germany with Winthrop Bell, who after being released as a prisoner of war had become the German correspondent for an English newspaper. It was Courant's contention that the intellectual middle class must involve itself in the struggle to rebuild Germany.

On a Sunday morning in January 1919, Courant gave a speech at an open meeting of the Social Democratic Party. It was entitled "Social Democracy, Revolution and National Convention" and was reported at length in a local newspaper.

At the time of the truce of October 1918, he pointed out to his audience, "if one had asked the German people which party they thought would save them from catastrophe, the overwhelming majority would have answered the Social Democratic." In a few months, though, the country's mood had changed. It was no longer so much in favor of the party, mainly —in his opinion—because of the short memory of the masses and the fact that right-wing parties were agitating, successfully it seemed, to make the Social Democratic Party responsible for the defeat of Germany and all that had followed.

Courant wanted to rebut these right-wing charges. If the debacle had been caused by any single thing, he told his listeners, it had been caused by "the egotistical blind attitude of the leading segment of the population." The revolution which had followed the armistice had not been instigated by the Social Democrats "nor by anyone at all," including the revolutionaries, "who overestimate their influence in this direction." It had come like an elemental force of nature. Once it had come, the Social Democrats had taken leadership, much less in the interest of their party than in the interest of the country as a whole, "to avoid inevitable chaos and civil war." The fact that the Reichstag was soon to meet again was a mark of the party's success.

Many interests were in conflict. He urged "a human understanding," not only of the middle class and the right-wing parties, but also of the radical elements of the working class.

"The democracies of the world, powerful in number, have thrown us to the ground," he concluded, "but under the sign of Socialism, the German spirit can reconquer the world!"

In addition to university and political affairs, there were personal matters demanding Courant's attention.

He and Nina were aware that a number of people in Göttingen looked with disapproval upon the fact that the daughter of a professor and the granddaughter of Emil DuBois-Reymond, the distinguished physiologist and philosopher of science, was planning to marry "a Semite." There were also Jews who disapproved of the match. "The Runges were such *decent* people," the Jewish wife of a Jewish professor said to me. In her view, which seems to have been the common one, a poor but ambitious young man had married a homely girl with a rich father and then, discovering in Göttingen how unsuited she was to further his academic career and seeing an opportunity to marry instead a famous professor's daughter, had dispensed with her.

Nina, who since childhood had said that she intended to marry a Jew —she liked their looks—and who knew that it was she who had proposed the marriage, was unaffected by the talk.

"Nina was always a kind of royal person," her friend Elli Husserl, now Rosenberg, told me. "Her mother and her aunts and uncles had played with the Hohenzollern children on the grounds of the palace, because old Emil DuBois-Reymond was a bigwig, one of the few professors—there were always a couple—selected to be socially acceptable at the court. So the world looked different to Nina, as it does when people are really free and independent and don't have to worry about how this or that might affect a relationship. I think the way that Nina proposed to Courant had something to do with this freedom she was born into."

It was different for Courant. He had been especially sensitive to the disapproval which seemed to him implicit in the fact that the Runges had put off announcing the engagement of their daughter for some time. He realized, he had written to Nina, that he was not the most desirable prospective son-in-law. He was a Jew "from a not especially good family," was also a divorced man, and was not yet well established in his career. Still, they *were* engaged, and her parents should tell people so.

Courant's parents were not happy about the match either. Siegmund, who in 1919 was bringing his memoirs to a conclusion, noted that Richard had not consulted him and Martha in the matter, just as he had not consulted them when he had married Nelly Neumann.

"It is not Mother's wish," Siegmund wrote of the alliance, "but Richard's will."

Since Courant and Nina were of different faiths, they could not have a religious service without one of them becoming a convert to the other's religion. "Many Jews converted in Germany," Nina explained to me. "It was often for 'political' reasons. I didn't want Richard to do that, and he didn't want to do it." They were married before a magistrate in the town hall on January 22, 1919.

"Now Richard is marrying Nerina Runge," Siegmund Courant wrote —it is the last line of his memoirs. "We shall see how it works out."

That spring, the new son-in-law of Professor Runge began to lecture as a *Privatdozent*.

The number of students at the university had increased dramatically. By governmental fiat many young men, including some who had not completed the last two gymnasium years, had been awarded the *Abitur*. Others who had obtained the degree in the normal way before August 1914 had, in five years of war, forgotten much of the mathematics they had learned. Special review courses were necessary, and the responsibility for these was delegated to Courant by Klein. Courant also gave other lectures on partial differential equations and mathematical physics.

The relation of the partial differential equations of pure mathematics to the problems which arise in physics was to fascinate and absorb him throughout his career as a mathematician. In that first post-war spring and summer, it was of special significance. The book on the methods of mathematical physics—the collaboration he had proposed to Hilbert— was much on his mind. He was also occupied with the development of a theory of the eigenvalues of partial differential equations, a subject intimately connected with the vibration problem of physics.

A student who heard Courant lecture at this time was Helmut Hasse. He had just completed the gymnasium when the war began. Serving in Kiel as a cryptographer, he had continued his education by attending a night class on the distribution of primes, which was taught by Toeplitz, a professor at the university there.

"That course was very decisive for my whole later development," Hasse told me when I talked to him in the spring of 1971 in San Diego, where he was lecturing for a semester at the state university on his now classical work in algebraic number theory. "Toeplitz made me familiar not only with the whole Hilbert spirit but also with the things he himself had seen and done when he was in Göttingen."

When Hasse could no longer attend classes because of his naval duties, Toeplitz encouraged him to continue his mathematical work and mail papers back to him for comment and suggestion. As a result, at the end of the war, the young man felt he was sufficiently prepared to present himself at Göttingen as a student of mathematics.

Hasse told me that although he liked Courant personally, he did not care for his subject matter or his way of lecturing. Unlike Erich Hecke, who "presented his ideas like an artist, taking great care for refinement and balance," Courant often made mistakes. These did not seem to embarrass him. He would simply wipe them out and say, "Oh well, let's try it this way."

After Courant found out about Hasse's interest in number theory, he gave him a manuscript by Harald Bohr.

"It was a wonderful thing on the Riemann zeta function and on diophantine approximation. I worked it through and reported on it to Courant. He invited me to tea at his home one afternoon—think of it, a young student—I was only twenty-one—getting to come into such contact with a professor!"

That first post-war spring, at the suggestion of Klein, who thought it would be advantageous for the university to have a representative in local government, Courant ran for the town council and was elected. For a while, also at Klein's suggestion, he considered becoming a candidate for the state parliament but ultimately gave up the idea.

He appeared in print once again in connection with politics. In July 1919 a certain Mr. Mühlestein, who had developed during the course of the revolution from a majority socialist to an independent socialist to a communist, had delivered an inflammatory speech on the current situation of Germany. Courant wrote a long letter to the local paper taking Mühlestein to task and concluding: "If you really mean well by Germany, if you want to put in your oar with her, why don't you carry your torch of world revolution to the Allies?"

During the summer semester of 1919, Courant did a great deal of mathematical work. By the middle of August he had finished and submitted for publication a long paper on the theory of eigenvalues of partial differential equations. He had high hopes that it would reestablish him as a mathematician.

That fall Erich Hecke left Göttingen for the newly founded university in Hamburg. The chair of Klein was again vacant.

Both Klein and Hilbert were eager to have Courant succeed Hecke. They knew, however, that the ministry would not approve the faculty's proposing to call as a professor one of its own—a man who had taken his doctor's degree and had obtained his *Habilitation* there. They conceived a stratagem to get around the obstacle. Wilhelm Killing was retiring from his chair at Münster, and it was arranged to the satisfaction of all parties that Courant would be called to Münster as Killing's successor.

NINE

IN THE SPRING of 1920 the period of the Social Democratic Party's dominance in the post-war government of Germany was essentially at an end. In the elections that year, parties to the right and to the left both made great gains. Courant was relieved to have an excuse to resign from the town council. He had found its sessions "more boring than faculty meetings" and had regularly come home with the dry announcement that he "had successfully worked for the end of the debate." Leaving Nina and their new baby son in Göttingen, he went cheerfully to Münster at Easter (the beginning of the German academic year) for what he hoped would be only a very short stay at that university.

The work which he had done on the eigenvalues of partial differential equations appeared in the new *Mathematische Zeitschrift* in 1920. Like so much of his mathematics, it arose out of a physical situation, reflected his interest in and enthusiasm for the work of people he knew, and had a connection with Dirichlet's principle.

The story of the work goes back to a lecture given in Göttingen before the war by H. A. Lorentz. In this lecture Lorentz referred to an interesting conjecture about the distribution of the eigenfrequencies of vibrating homogenous media, which correspond to the fundamental tone and the succession of overtones of acoustical systems. Hermann Weyl immediately began to work on the related mathematical questions; and in a series of powerful and ingenious papers, which he began to publish even before the war, he verified the Lorentz conjecture for a number of the most important vibration problems.

Courant was tremendously impressed by Weyl's work, but he felt that it lacked esthetic appeal. The method was "roundabout" and therefore did not give a really complete insight into the relationships involved. After his own work on Rayleigh's theorem during the war, he thought he saw a way in which he could verify the Lorentz conjecture more directly than Weyl had.

To get a glimpse of Courant's method and its relation to that of Weyl, one must go back to the facts on which Lorentz based his conjecture. At the time it was already known that the eigenfrequencies under consideration depend in a rather complicated manner upon the nature of the shape of the vibrating medium but that they can be described in a simple way, which becomes increasingly accurate as the frequencies become higher. It was

Lorentz's conjecture that the distribution of very large frequencies depends, not on the shape of the vibrating medium, but only on its volume.

The behavior of the eigenvibrations is described mathematically by a partial differential equation which has a solution only for special values of a certain parameter. These are the eigenvalues of the equation, the squares of the eigenfrequencies. At the time Lorentz offered his conjecture, mathematicians considered the partial differential equations which describe the behavior of the eigenfrequencies almost impossible to handle, since there were many severe technicalities connected with them. For this reason Weyl transformed the partial differential equation into an integral equation and then employed the theory of integral equations to verify the Lorentz conjecture.

"But Courant hated the method of integral equations," Friedrichs explained to me. "He thought one should be able to work directly with the relevant partial differential equations—that there was no necessity to transform them into integral equations as Weyl had done in his work. Technicalities that Weyl avoided by his approach didn't bother Courant so much. He was always inclined to feel that such things could be managed one way or another—even if he didn't know quite how at the time. And in the end, as it happened, he *was* able to straighten them out."

Courant wanted to show—as he wrote in the long paper published in 1920—that "from another point of view one can gain direct access to the whole complex of questions posed and can attain complete command of the eigenvalues . . . , the boundary conditions, and the domain in a surprisingly simple and consistent way . . . which, in economy, essentially surpasses Weyl's method and seems to be theoretically satisfying."

The leading notion of the method which Courant proposed was to characterize the eigenvalues as the extreme values of certain integrals connected with the original partial differential equation of the vibration problem. The method was of course a close analogue of the Dirichlet principle.

Even before the work of Weyl, it was known that the eigenvalues of the related partial differential equation could be characterized as those values for which a certain equation has a solution other than zero and that this solution—the eigenfunction—gives the amplitude of the vibration. It was also known that the lowest eigenvalue could be characterized by a minimum property closely related to the Dirichlet principle and that every eigenvalue could be characterized by a minimum property once all the previous eigenvalues and eigenfunctions were known. Although theoretically satisfying, this method was impractical to use in the case of very high eigenvalues. Courant, however, observed that it is possible to characterize any eigenvalue by a modified minimum property *which does not depend on the knowledge of the lower eigenvalues and eigenfunctions.*

"You can start right in there!" Friedrichs told me with enthusiasm. "Oh, you have to satisfy some additional conditions, of course; but the important thing is that you don't have *to know* the previous eigenfunctions. The conditions, the additional conditions, are of a very general character. You see, Courant's observation was the following. If you impose certain orthogonality conditions on the appropriate number of arbitrary functions and then take the minimum, that's not good enough; but if you then *maximize the minimum,* it comes out right. That gave him the possibility of direct characterization of the eigenvalues—the squares of the eigen-frequencies—and it was this that permitted him to get the asymptotic behavior of the high frequencies of the Lorentz conjecture and of the eigenvalues of the associated partial differential equations. It was a beautiful idea. *A very beautiful idea."*

As it happened, unknown to Courant at that time, this same maximum-minimum principle had already been stated by Ernst Fischer and used for much simpler cases having no connection with the Lorentz conjecture or with mathematical physics. But, in Friedrichs's view, the important thing was not so much to have discovered the principle as to have seen that the reformulation would allow direct approach to the problem of asymptotic distribution.

"That, I think—the twist, *that* was Courant's."

Both the method of Courant and the method of Weyl had come out of Hilbert's work at the beginning of the century. At the time of the publication of Courant's work in 1920, it still seemed to most people that while his method might be more esthetically satisfying, Weyl's method—which was in some ways more specific than the method of the Dirichlet principle—was the more powerful. In the long run, according to Friedrichs, Courant's method seems to have proved superior to that of Weyl. But that was not clear at the time when Courant was a young professor in Münster trying to reestablish himself as a mathematician.

Courant's appointment at Münster was as a professor of mathematics; but, according to his contract, he also seems to have delivered lectures on the applications of mathematics as well. This was probably the reason that during the term he was approached by a middle-aged man with the kindly, open face which he always associated with peasants. Conversation revealed that the visitor, whose name was Carl Still, had studied on his own such writers as Lagrange and Clausius—with remarkable tenacity, in Courant's opinion—but now felt that he needed more than he had been able to get for himself from these authors. He talked with Courant about certain problems—"non-trivial" to the professor's surprise. Some were purely theoretical; others were important to Still's work and had to do with such applications as the determination of flow conditions by calculation and the material and energy balances for chemical processes.

The mathematical discussions led eventually to an invitation to Courant to visit the Still home. To his amazement he discovered that his simple visitor was in fact a well-known self-made industrialist of the Ruhr area.

As far as I know, the beginning of this friendship with Still was the only significant event of Courant's stay at Münster. Even before the end of his first semester there, Klein and Hilbert had begun to push openly in the faculty for his return to Göttingen.

Obtaining a professorial appointment for Courant would have been a relatively simple thing if it had not been for the fact that, at the same time, Klein and Hilbert were proposing to add *two* new physicists to the faculty to take the place of Peter Debye. Originally, Courant told me, he himself had suggested Born to Hilbert as a replacement for Debye. Born had been happy to return to Göttingen but had felt a little overwhelmed by the assignment. Although he had had experience in experimental as well as theoretical research, he did not feel that he was capable of directing the big physics laboratory, which in his student days had been divided between Voigt and Riecke. A little research on his part revealed that, although one physics professor, Robert Pohl, had recently been appointed, there was still a place on the university books for yet another physics professor. On the basis of this fact, Born made it a condition of his acceptance that his friend James Franck, an outstanding experimental physicist, also be called to Göttingen.

"Franck + Born are the best imaginable replacement for Debye!" Hilbert wrote delightedly to Courant in Münster. "I am very happy about this arrangement. We have Born's energy to thank for it!"

Thanks to Born's energy, however, Klein and Hilbert found themselves pushing for three mathematical-physical appointments at the same time. There was some opposition in the faculty—"because of a little hostility toward me personally," Hilbert informed Courant. One professor suggested that the faculty consult outside experts before making its decision. Landau proposed as a compromise that only Courant be called at that time.

Undaunted, Hilbert wrote cheerfully, "We have much reason to be optimistic. Klein holds out bravely!"

At the same time Klein also wrote to Courant, "As you may have heard from other sources, I intend to advocate your appointment in Göttingen. It would be extremely helpful for me if you would confirm explicitly in writing that you are willing to promote with energy tasks which, in my opinion, have long been unduly neglected in our educational system and new demands which I can foresee as coming up."

He then listed three items: (1) regular introductory lectures for students not headed for abstract mathematical training, (2) the taking over of the organizational duties which he himself had handled over the years, and (3) "a positive orientation" to all important questions arising out of the revamping of the educational system at the university level as a result of the post-war situation.

He noted in a postscript: "Not one of these three points should come as a surprise. . . ."

"Klein has read me his letter to you," Hilbert explained to Courant, "and he wants your reply only to be able to assure the minister that he has your written support for his organizational efforts."

Klein was old, ill, and tired; but the rest of the faculty was no match for him. All three appointments were approved.

Born's and Franck's assumption of their duties was delayed by the fact that they couldn't obtain housing in the crowded post-war conditions of the university town. Since Courant had not given up his apartment when he left for Münster, he was able to return to Göttingen immediately.

For an American, accustomed to the American university system of separate departments, each with its own chairman, a budget and a building of its own, it is hard to conceive of the undefined nature of Courant's role as Klein's successor.

Klein had never held a formal administrative position. His fully autonomous colleagues had simply been willing to let him handle what Hilbert dismissed as "the mathematical arrangements." It is true, as Weyl later wrote, that Klein ruled in Göttingen like a god, but his godlike power came from the force of his personality, his dedication and willingness to work, and his ability to get things done. Over the years he had concerned himself with the reform of secondary education in the sciences; the improvement of the technical schools; the development of comprehensive mathematics courses for teachers and engineers; the securing of the connections between mathematics and other sciences and between mathematics and its applications; the establishment of close relations between industry and science.

In Göttingen in 1920 there was no mathematics department. There were four mathematics professors who, like the other scientists, were members of the Philosophical Faculty, which included philosophers, historians, philologists, and classicists, among others. There were also some private lecturers in mathematics. There was no budget, except for the running of the library; and there was no building—in fact, the professors did not even have offices.

Klein had dreamed of a single building—an "institute," such as those of the experimental sciences—which would house the *Lesezimmer* and all the other mathematical activities. He had desired above all else to make Göttingen the mathematical-physical center of the world. Courant was personally very much in sympathy with Klein's ideas; and he thought—he told me—that in his own way he might be able "to move them a little bit forward."

Throughout Klein's career he had had strong allies in the non-academic world. Among these was the publisher B. G. Teubner in Leipzig. But even before the end of the war Teubner had been showing himself unwilling to provide the support to mathematics he had given in the past. Like most German publishers, he was eager to pare down his operations "until normal times returned." He had let it be known that he would welcome being relieved of the *Mathematische Annalen.* He had further scandalized everyone in Göttingen by suggesting that the paper stock on hand for the collected works of Gauss—still not yet all in print sixty-five years after the death of the prince of mathematicians—should be used for other, more profitable purposes. In these circumstances Courant wasted no time in introducing his new friend Ferdinand Springer to Göttingen.

Unlike his fellow publishers, Springer recognized "that the time of aberration from 'the normal' had begun" and it would be a long time before it ended, if it ever did. Even before the war was over, he had founded the *Mathematische Zeitschrift*; and at the end of 1918 he had contracted with Courant for the *Grundlehren* series. In 1919, while Germans were expressing their outrage at the peace terms offered at Versailles, he took over the distribution of the works of Gauss from Teubner. In 1920, as the

Weimar coalition of the Social Democrats began to fall apart, he signed a contract with Teubner to become the publisher of the *Mathematische Annalen*. In 1921, as the Allies threatened occupation of the Ruhr, he published the first volume of Courant's yellow-covered *Grundlehren* series; and in the next three years, which saw the worst of the post-war inflation, he brought out a dozen more volumes.

The friendship of publisher and scientist was mutually rewarding. Springer took Courant's advice in entering the new fields of mathematics and physics. Courant made every effort "to protect my friend Springer from risks and losses."

As part of his contract with the ministry, Courant had negotiated a generous provision for two personal assistants. His first assistant was Hellmuth Kneser, who was getting his Ph.D. with Hilbert on a topic in the foundations of quantum mechanics. Kneser, who became much better known as a topologist, was the son of Adolf Kneser, who had been one of Courant's teachers in Breslau. In later years, when his own son Martin Kneser had also become a mathematician, he was to complain humorously that he had spent the first half of his life as the son of Adolf Kneser and the last half as the father of Martin Kneser. As Courant's assistant from 1920 to 1924, Kneser did the usual things to help the professor and also helped younger students in many ways. B. L. van der Waerden remembers how, as a young doctor from Amsterdam, he regularly lunched with Kneser and then took a mathematical walk, "on which I always learned a great deal."

While settling into the academic life in the winter of 1920–21, Courant became politically active again for a brief time. A plebiscite was to be held in Upper Silesia to determine whether that area, which included his birthplace, would become Polish or remain German. The right to vote had been given to all persons born in the area who would have completed their twentieth year by January 1, 1921, regardless of their current residence. The German government mounted a massive campaign to bring back as many as possible of those who had left the area to vote in the plebiscite. Courant took an active part, writing to friends and relatives he had not seen in years. On the day of the voting he was among the more than sixty Courants who gathered in Lublinitz.

Lublinitz itself recorded a German majority in the plebiscite, but it was included in that part of Upper Silesia that went to Poland. The Courants who had remained there, running the business of "S. Courant," chose to leave their birthplace forever.

After the plebiscite of 1921 Courant continued to maintain a lively interest in political affairs, but he devoted himself personally only to those aspects which directly concerned science and the university.

By the beginning of the winter semester 1920–21, James Franck had been able to establish himself as professor of experimental physics; and at the beginning of the summer semester 1921, Max Born arrived to take over his duties as professor of theoretical physics.

Born was an exemplar of what Courant considered "the Göttingen tradition." He had come originally to the university because it was "the mecca of mathematics." Then, as the tale is told, Hilbert had given him a problem which he was not able to solve—a problem still today unsolved—and Born, doubting his mathematical abilities, had turned to physics. By the time he came to Göttingen as a professor, he had done his famous work on the derivation of all crystal properties from the assumption of a lattice whose particles could be displaced under the action of internal forces; developed with Fritz Haber a heat theory which included the first example of the determination of a chemical heat reaction from purely physical data; and tried his hand at experimental research, discovering with the help of his assistant E. Bormann a method for determining the free path of a beam of silver atoms in the air.

James Franck had already produced the work on the changes of energy occurring upon the collision of atoms with electrons, for which, in 1925, he would receive a Nobel Prize.

During the year 1921 two outstanding post-doctoral students also arrived at the university.

One of these was Carl Ludwig Siegel, who had been encouraged by Landau to come to Göttingen after the war and had then gone with Hecke to Hamburg. Siegel's mathematical interest was different from Courant's, and the two men had not met during Siegel's initial stay in Göttingen. Courant, however, had heard from others of Siegel's abilities and of his current situation in Hamburg, where he was cold, hungry, and unhappy.

At the beginning of 1921, Courant wrote to the ministry suggesting that the second assistantship provided for in his contract should now be activated and offered to Siegel.

It did not take any special perspicacity on Courant's part to see that Siegel was going to be an outstanding mathematician. What Courant did see was that Siegel was a young man, "not easy," who required and should receive special treatment because he was "an absolutely unique talent." He managed to convey this to the minister.

"It would make me very happy," the minister wrote to Courant, "if this grant would suffice to support so powerful a talent as, according to your presentation, Mr. Siegel is, and enable him to continue his scientific career."

When Siegel returned to Göttingen, the little Courant paternally took charge of the tall youth, found a place for him to live in a vacant room at Klein's house, invited him to swim at the faculty bathing establishment, saw that he met Hilbert there and that Hilbert learned about his work, and invited him frequently to his home. There Siegel, much to the amusement of Nina, sat on the floor and solemnly tried to get baby Ernst to repeat after him long and complicated scientific terms.

Courant also made an effort to draw Emil Artin, the other outstanding newcomer, into the inner circle of Göttingen. Artin was an Austrian with talents and interests in art and music as well as mathematics. After serving in the war, he had gone first to Vienna to study, then to Leipzig. He had taken his degree with Gustav Herglotz, who had been an associate professor in Göttingen during Courant's student days. It was Herglotz's suggestion that Artin go to Göttingen for post-doctoral study. Although, as in the case of Siegel, the mathematical interests of Artin were quite different from his own, Courant welcomed the young visitor, invited him for musical evenings at his home (where Artin, whose music was as pure and rigorous as his mathematics, shuddered at Courant's untutored approach to the keyboard), and saw that Klein and Hilbert met him and learned about his work.

This maintaining of contact between the new young people and the great men of the past was a conscious effort on Courant's part. He felt that Göttingen's mathematical-physical tradition of teaching combined with research had been almost completely the personal creation of Klein. Gauss had had little connection with the instructional side of the university and considerable distaste for it. Dirichlet had begun to give advanced courses, but these had not drawn a large number of students and the development had been cut short by Dirichlet's death, four years after that of Gauss. The sick and shy Riemann had not had the ability to carry on what Dirichlet had begun. It was Klein who had taken these great names and their wide-ranging research activity and molded them into a tradition.

Now, in the first post-war years, Courant began to weave into the Göttingen tradition the great living figures of Hilbert and Klein.

TEN

SHORTLY AFTER Courant returned to Göttingen, he introduced his friend from Münster, the industrialist Carl Still, to the Göttingen circle. At the begining of 1922, Still contributed 100,000 marks to the funds remaining in the treasury of the *Göttinger Vereinigung*. A new organization was formed with the same purposes as Klein's old group—the cultivation of the applications of mathematics in Göttingen.

That same year the German mark, which had had a pre-war value in relation to the American dollar of 4.20 to 1, fell from 162 to 1 to 7000 to 1. It was clear that imagination was going to be as necessary as money for mathematics in post-war Göttingen.

In spite of the rapidly increasing inflation, the year 1922 was to see important milestones on the road to recovery for mathematics and physics in Göttingen and in Germany as a whole. It began with a great celebration of Hilbert's sixtieth birthday. It saw, in the summer at Leipzig, the first general meeting of German scientists since the war. In the fall, in Göttingen, there was a historic week-long series of lectures in which Niels Bohr explained for the first time in public the strange new ideas which so many had rejected before the war and which were, at the end of 1922, to earn him the Nobel Prize for physics.

When, that same year, mathematics and the natural sciences were formally separated from the other specialties of the Philosophical Faculty, Courant applied for permission to change the name on the stationery which he and the other mathematics professors used for official business from *Universität Göttingen* to *Mathematisches Institut der Universität*. This seemingly innocuous request was granted.

Supported by Klein and Hilbert, Courant tried to lure back to the university the outstanding German mathematician of his generation—the young man whom Toeplitz had once identified as someone else who also thought about mathematics.

Hermann Weyl had left Göttingen in 1913 with the beautiful Hella as his wife and, after service in the army, had become a professor at the Eidgenössische Technische Hochschule in Zurich. There, for a year, he had been in close contact with Einstein. After Einstein had presented his general theory of relativity in November 1915, Weyl had delivered a popular series of lectures on Einstein's ideas, later published as *Space, Time, and Matter*. Courant was among those who had immediately recognized the significance of Weyl's mathematical-physical-philosophical orientation for the new situation in science resulting from Einstein's theory. In a letter to Hilbert in July 1918, he had suggested that the mathematician Weyl should be called to Göttingen to succeed the philosopher Husserl, who had received a long overdue professorship in Freiburg.

"The department of philosophy at the universities today is conceived as if philosophy as a science in which fixed results can be formulated does not exist . . . ," Courant had written to Hilbert. "It has come to the point that a mathematician or a physicist rather fears the investigation of his own fields by philosophers and no longer expects enlightenment or even understanding from them. . . . Now, after Einstein's colossal achievement, the deep significance of which lies particularly in the philosophical area, . . . this situation must come to an end. . . . It is definitely desirable that a number of philosophical positions be held by people who are closer to mathematical and scientific thinking. Since such people are not easily found among the regular philosophers [with the exception of Husserl and Nelson], a mathematician like Weyl would be suitable for such a position," he had suggested.

Husserl's place was filled, with Hilbert's backing, by Leonard Nelson; but Courant continued to hold the idea that if such an appointment as that of Weyl was not made, the old "unfriendly" conditions between mathematics and philosophy would continue to exist. In 1922, with the establishment of a separate faculty of mathematics and the natural sciences, Courant apparently negotiated an additional chair for mathematics and offered it to Weyl.

Weyl was torn by the opportunity to return to Göttingen. On one side were the attractions of the German university to which he was always to considered himself inextricably bound and to which he often returned for the stimulation which he had received in his youth from Hilbert and Minkowski; on the other side, the incontrovertible uncertainty of the conditions of life in Germany—the new republic battered from right and left, constant threat of revolution and continuing acts of terrorism by extreme nationalistic parties, inflation now "galloping."

After a long delay and much debate with his wife, he finally set off for the telegraph office with his acceptance written out and in his hand. By the time he arrived, he had changed his mind. He wired instead his refusal.

The economically and politically miserable year of 1922 which had deterred Weyl from returning to Göttingen saw the arrival there of a number of gifted students. One of these was Friedrichs, then an awkward, quiet, and asthmatic youth, very shy.

He found the mathematical life in Göttingen "a terrific shock." The level of the subject matter was far above that to which he had been accustomed. A seminar conducted by Courant was based on a book which he had already encountered at another university, "but there, during the whole term, we had gone through chapters one and two—in Göttingen we started at chapter three."

He also found the subject of Courant's seminar an unusual one. Since his dissertation Courant had been interested in algebraic surfaces. There were currently in Göttingen two young men he much admired—Siegel, now a *Privatdozent*, and Artin; and both were well informed in algebraic number theory. Although the two topics shared the name "algebraic," they were not as a rule combined; but Courant decided there should be a seminar taking up algebraic surfaces one week and algebraic number theory the next week.

The first report in the seminar of 1922 was given by Otto Neugebauer. He was also an Austrian and had served as an artillery officer during the war. In studies in Graz and Munich he had moved from electrical engineering to physics to mathematics. From the moment that Courant heard Neugebauer's seminar report, he was completely enchanted by the young man and was soon on friendly, almost colleagual terms with him.

Friedrichs remembers the seminar as being impressively well attended.

"Siegel and Artin came, also Kneser. There was present every assistant who was in Göttingen at that time. They all had to come and participate. Such a group of people, who knew everything about everything—it was very exciting to me."

"When you say 'they had to come and participate,' do you mean that Courant required them to come, or that it was customary for everybody to come to the seminars?"

Friedrichs laughed.

"That question cannot be answered. Such a notion did not exist for Courant. He did not ask the people as a requirement as Klein might have done. No. He would say, 'It's very important that you help us with this, we need your help'; and he would manage somehow that everybody who had something to contribute did attend. That's the way he always operated."

While Friedrichs was being awed by Siegel, Artin, Kneser, and the other young mathematicians of Göttingen, he himself was putting in awe another student. Hans Lewy was several years younger than Friedrichs. He was straight out of the gymnasium, a small, bright, extremely sensitive and impatient boy, musically as well as mathematically gifted. In Göttingen for the first time in his life, he told me, he found himself confronted by a number of people who seemed to know much more than he did. He was especially impressed by Friedrichs.

When I talked to Lewy in Berkeley, where he had recently retired as a professor at the University of California, I asked him how he had liked Courant as a teacher in those days.

"Courant was, for me, a very good teacher," he replied. "Of course, often his lectures were not sufficiently prepared, and maybe he would have to change his approach during the hour because he had noticed that something was missing. But this, in some ways, was stimulating. For one thing it allowed him to say much more, to allude to subjects which he was not fully covering, to stimulate the ambition of the students to cover the gaps, to see clearer."

To Lewy it was especially important that Courant seemed to be able to communicate a lack of fear of the technical details which often overwhelm students.

"Courant had the attitude that the technical details would take care of themselves. That inspired young men. It is especially important in analysis because, as analysis is a very old subject and a vast subject, it takes a long time to get to the point where one understands where the problems lie. That, in my opinion, is partially the reason that analysis has little success in the United States and little attraction for the young. Courant was able to make a young man feel that one could just break through. . . ."

"Yes," he concluded thoughtfully. "That was a great help."

By the time that Friedrichs and Lewy arrived in Göttingen in 1922, three of the monographs in Courant's Yellow Series were in print or in the process of publication. But his own book, the collaboration with Hilbert on the methods of mathematical physics, was still not finished.

His original plan had been merely to edit the notes of Hilbert's lectures, which he and other assistants had written up before the war. Then gradually he had begun to add things of his own and to rearrange ideas. Certain areas which did not appeal to him were dropped. Other areas which he liked very much were given an increasing amount of space. In particular, he wanted to elaborate on the theory of eigenvalues which he had developed after the war and the related theory of eigenfunctions on which he had worked since then. He felt that he had first to lay a foundation for his ideas with an extensive presentation of a number of methods suited for many different special cases. This material was already in existence but was scattered and not in easily accessible form.

Courant obtained funds, perhaps from Still, for a *Hilfsassistent*, "an assistant to the assistant," to take notes on his own lectures on partial differential equations. Several students were eager to have the job, and he had them write up the first lecture of the semester as a sample of what they could do. He ultimately selected Pascual Jordan, a physics student who had prepared a set of notes although he had not heard the lecture. Jordan also helped Courant with the preparation of the book on the methods of mathematical physics.

As the possibility of publication of the book with Hilbert continued to recede, Courant saw an opportunity to produce rather quickly a collaboration with another famous mathematician on a subject which also had great attraction for him. Adolf Hurwitz had died at the end of 1919. Although Courant had not been inspired by Hurwitz's lectures as he had been by Hilbert's, he felt strongly that they should be made available to a wider audience than Hurwitz's own students.

The Hurwitz lectures on function theory were in the spirit of Weierstrass. They were straightforward, precise, remarkable for their clarity and their esthetic qualities. Courant admired them greatly, but he did not think that they hit "quite the center of interest." He felt a need to balance them with the approach of Riemann which was—in contrast—geometrical rather than arithmetical and intuitive rather than logical and which drew its inspiration from the problems of physics rather than from those of pure mathematics.

The Hurwitz-Courant book on function theory appeared before the end of 1922. Reactions to it were mixed.

The American mathematician Oliver Kellogg was especially critical. He found the Hurwitz lectures "lucid" and "noticed no faults in logic" in them; but as to Courant's contribution—"It gives the impression of being the work of a mind endowed with fine intuitive faculties, but lacking in the self-discipline and critical sense which beget confidence. . . . What is found may, indeed, serve as an indication of some of the directions which modern investigations have taken—in fact, a very interesting one. But the proofs offered often leave the reader unconvinced as to their validity and, at times, uncertain [even] as to whether they can be made valid."

A different reaction to Hurwitz-Courant was that of Friedrichs, who read the book as a student at about the same time that Kellogg was reviewing it. To Friedrichs the first two sections by Hurwitz seemed "very neat, very clear, to the point, you could learn from them, but they were not inspiring. But the Courant section, the third chapter—when I got hold of that chapter, I started reading one morning, I read morning and night without stopping. It was the most breathtaking book I have ever read in mathematics."

(The first edition of Hurwitz-Courant sold out within two years. Courant then asked his young friend Neugebauer to eliminate the flaws in precision and logic which Kellogg had criticized. Neugebauer, whom Courant once described as "having all the virtues of pedantry and none of the vices," did an excellent job. Kellogg in a second review noted the improvement. But for some readers a little of the magic of the first edition had been lost.)

The Courant approach to function theory, so repugnant to Kellogg and many others, was that of "the romantic type" in the sciences as opposed to "the classic type." This distinction was originally made by the chemist Wilhelm Ostwald. Courant once cited it in describing Klein, whom he saw as an example of the romantic type:

"While the classic type in the sciences carefully examines every detail and repeatedly refines and polishes his work before permitting it to leave the seclusion of his study, the romantic type throws his discoveries to the public as an immense stimulus, often before his ideas have reached full maturity. The classic type prefers to lock up in his desk three quarters of his scientific output if a minor point does not satisfy him, and he never wishes to say more than he will be able to support in years to come. The romantic type, on the other hand, does not place great value on the fully matured and completed form, and he does not feel abashed if he has once said more than he can actually prove. He is interested in the immediate, vital impact. He lives within a circle of enthusiastic disciples who are greatly enriched by their contact with him and who will one day move on to do their own work."

In 1922, in addition to publishing the book on function theory, Courant also published his theory of eigenfunctions, which had grown out of his work on eigenvalues. In the earlier work, he had simply assumed the existence of the eigenfunctions associated with the eigenvalues. Now he successfully tackled the problem of proving existence. In the course of this work he used techniques that differed in some ways from those he had used before and introduced a number of elegant geometric-algebraic concepts. Today other, though related, concepts developed in functional analysis are considered more appropriate for the purpose. "But Courant was always very slow in accepting notions that grew out of this much more abstract field," Friedrichs told me. The work on eigenfunctions was not to have as great an effect as the earlier work on eigenvalues.

The paper of 1922 on eigenfunctions did, however, initiate a series of papers on the partial differential equations of mathematical physics in which Courant's primary concern was *existence*. The significance of this concern on the part of mathematicians is sometimes questioned by even quite sophisticated physicists. They are inclined to feel that if a mathematical equation represents a physical situation, which quite obviously exists, the equation must then of necessity have a solution.

Courant, however—as Friedrichs said—"simply refused to be intimidated by the objections of such physicists." He was convinced that existence investigations would contribute to the understanding of the nature of the equations and their solutions. He also felt that a knowledge of existence considerations would be helpful in setting up schemes for numerical computations. From 1922 on, he emphasized questions of existence in all his work on partial differential equations of mathematical physics.

The year 1922, which was one of the most mathematically active of Courant's career, was also a year in which shelter and food were beginning to be more immediate problems than those of mathematics. It was natural for students to turn to Courant for help, and he often turned to Carl Still.

Still was the kind of successful man of business or industry whom Courant was always to admire extravagantly. Although he was "very hard and very sharp," the more important requisites of his financial success were for Courant "expert knowledge and experience, diligence, and uncompromising adherence to the highest ethical standards of business life." Courant also saw in both Still and his wife, Hanna, a deep-rooted devotion to that which transcends mere personal values—"a *religious* quality in the real sense of the word."

If 1922 was economically miserable, 1923 was immeasurably worse. Rents were sometimes set in terms of butter, which was virtually unobtainable; and potatoes were on occasion purchased with baskets and later with wheelbarrows of currency. Money sent from home lost in purchasing power before it could arrive and then again before it could be spent. Many students worked on the railroad, shoveling gravel back under the ties, in a futile effort to supplement their incomes; but, as Lewy told me, "it was impossible to have *enough* money." Sitting before their books in the *Lesezimmer*, they could not get their minds off food.

Yet for the Göttingen mathematicians there were also hopeful signs in 1923 of a return to contact with the world outside Germany.

In that spring two Russian mathematicians arrived, the first scientists to be "sent out" of Russia since the Bolshevik revolution. P. S. Alexandroff and P. S. Urysohn were products of the post-revolutionary school of Luzin, but already they were on their way to founding their own new school of topology.

The same year which saw the arrival of these first visitors from the east saw also an opportunity offered from the west for young German mathematicians to go abroad to study once again. In 1923 the International Education Board, newly founded by John D. Rockefeller, Jr., with Abraham Flexner as Director of Educational Studies, announced a series of fellowships to assist young scientists to pursue in other countries "studies which they cannot pursue at home with equal advantage." There were places where young German mathematicians would still not be welcome; and for the first in what was to be a long series of recommendations, Courant turned to the greatest foreign friends of Göttingen—Harald Bohr and G. H. Hardy.

By 1923 there was no fixed currency in Germany. In a talk many years later, Neugebauer vividly recalled the situation as follows. Salaries and prices were expressed in basic numerical classes which were then multiplied by a rapidly increasing coefficient $c(t)$ such that the result represented their value expressed in marks at a given time. On a certain day of the week the current value of $c(t)$ was divulged to the business office of the university, hours before it was announced in the press.

Recognizing that early knowledge of $c(t)$ would greatly increase the purchasing power of funds allotted to the *Lesezimmer*, Courant offered to lend the business office the "mathematics institute's" electric calculating machine. In return he would be informed of the value of $c(t)$ as soon as it was received. Inflation was proceeding at such a rate that on July 1, 1923, the mark stood in relation to the dollar at 160,000 to 1; and on October 1, at 242,000,000 to 1. Courant's machine had a range of 19 digits. The university business office did not hesitate even an instant in accepting his offer.

By November 20 the mark has a value 4,200,000,000,000 to the dollar. The government announced that a state of national emergency existed. A new currency was introduced in strictly limited quantities and backed by a mortgage on all the industrial and agricultural resources of the country. From one day to the next, the currency was amazingly stabilized. A few months later, the government was able to declare that the state of national emergency was officially at an end.

It was during this period that Courant completed *Methoden der mathematischen Physik*—the book which he had proposed in 1918 as a collaboration with Hilbert. Since then, Hilbert's health had progressively failed. His mathematical interest—with characteristic completeness—had changed from physics to the foundations of mathematics. He showed an interest in the book his former student was writing but did not participate in any other way.

At one time, Friedrichs recalls, Courant showed him the notes of Hilbert's lectures, on which he had originally planned to base the book. These were beautiful, but they were not what is known as Courant-Hilbert. Friedrichs took his copy of the first volume out of his bookcase and leafed through it.

"It *is* a rather unsystematic book," he conceded. "Here, also, you see, Courant wanted to *combine*, like in the seminar on algebraic surfaces and algebraic numbers and in Hurwitz-Courant. There are various approaches to a theory of partial differential equations. Then he has an introductory section on algebraic notions, then some semi-algebraic methods, then some series expansion, then integral equations, then calculus of variations—and then he comes to vibrations—his own work on eigenvalues and eigenfunctions. Certain areas completely omitted, certain areas which Courant liked overemphasized. Very unsystematic."

The influence of Lord Rayleigh's *Theory of Sound* on Courant's book was apparent; the spirit of Hilbert "hung over every page" (as Paul Ewald noted). Reviewers recognized immediately, however, that in spite of the joint authorship, Courant-Hilbert was obviously written by Courant. A considerable portion of it was based on his own investigations. "Otherwise," pointed out the American mathematician Einar Hille, "the simple choice of methods, the fondness for heuristic considerations, and a certain delicate touch of pen, sometimes a bit vague but always elegant, betray the writer if nothing else does." The book also contained various errors, a number of which were to persist into the English translation made many years later. There—according to Alexander Weinstein, a later reviewer—"being by now classical, they only add to the pleasure of the reader."

The mathematical methods treated in Courant-Hilbert, as in Rayleigh, pertained to subjects which had originated in such areas of physics as elasticity, acoustics, hydrodynamics, and other classical subjects. In the hands of mathematicians these methods—which had their roots in physical intuition—had been transformed into rigorous tools backed by general theories. It was Courant's purpose to return the improved tools to the physicists for their own work.

At the time of its initial publication, however, Courant-Hilbert seemed of more interest to mathematicians than to physicists. The physics with which it concerned itself appeared to be, in 1924, rather old-fashioned to men who were striving toward some sort of understanding of the quantum of energy. In fact, Courant-Hilbert seemed a perfect illustration of Friedrichs's definition of applied mathematics as "those areas of physics in which physicists are no longer interested." But Courant was absolutely convinced that the content of his book would be important for physicists.

"You couldn't have told him different," Friedrichs said. "He wouldn't have believed you. Because of his optimism."

With a smile Friedrichs put his copy of Courant-Hilbert back into his bookcase.

"And of course he was right."

ELEVEN

THE POST-WAR YEARS saw a number of changes in Göttingen. As Klein's successor, Courant quite soon made an instructional innovation, the *Anfängerpraktikum* or "beginners' practice period," which was to have important results for the new "institute."

The *Praktikum* paralleled the calculus lectures. The students, often numbering as many as two hundred, received a mimeographed sheet of problems, some requiring inventive thinking as well as understanding of the material of the lectures. The professor regularly held a conference with a group of older students, discussing the problems and pointing out different methods of attack and various aspects of the solutions. The older students then went over the problems in the *Praktikum* with the beginning students and at the same time became personally acquainted with them. Solutions were written up and graded. Collusion was encouraged. Attendance was purely voluntary.

It is hard to realize today what a revolutionary innovation the *Praktikum* was in a German university at that time. Up until then, problems were never handed out except in applied courses, where they were usually not corrected. Textbooks were rarely utilized. Examinations were not given. The whole system was one of lecturing on the part of the professors and listening on the side of the students. The moment of truth did not arrive for several years, when the students had to take the state examination for teachers or the oral examination for the doctor's degree. For some, only then did it become clear that *knowing mathematics* is not like knowing the plot of a work of literature or the general outline of a historical period. The shock they experienced at this revelation not infrequently resulted in a nervous breakdown.

The *Praktikum* was a way of coping with a greatly increased number of students and a much lower level of ability and preparation. It required a group of older students to supervise it and these became additional "assistants" with appropriate financial support from the government. It also required *space*. As a result Courant was able to obtain rooms in an old building on Prinzenstrasse, from which Wilhelm Weber and Gauss had strung the wires of the first telegraph to Gauss's observatory. With this increased base, the paper "institute" was on its way to physical reality.

Klein never tried to interfere, Courant told me, but yielded the reins gracefully. There were, however, still signs of the old imperiousness. A favorite story of Courant's was how Klein, convinced at one point that death was actually at hand, called his assistant and dictated the arrangements for his funeral "and then was very displeased that he had not died after all." But there were also signs of mellowing. When a group of teacher-training students complained that Neugebauer, the assistant in charge of the *Lesezimmer*, had relegated Klein's collection of books on elementary mathematical pedagogy to the topmost shelves and packed them in so tightly that not a single one could be removed except with great difficulty, Klein summoned Neugebauer. He made only one comment to the young man, whose growing interest in Egyptology he was aware of.

"There came a new Moses," Klein said, "who knew not Pharaoh."

Changes were also taking place in the famous faculty.

Hilbert worked with passionate intensity on his program to shore up the logical foundations of mathematics against the attacks of L. E. J. Brouwer and his followers; but he was seriously, perhaps fatally, ill with pernicious anemia. In the spring of 1925 Runge retired; and Gustav Herglotz, a remarkably wide-ranging mathematician, the teacher of Artin, succeeded to Runge's chair. In June, that same year, Felix Klein died.

Courant wrote two moving articles about Klein. One was an account of his activities as a scientific leader; the other, an account of his life, work, and personality.

In later years, it is true, Courant was inclined to sacrifice Klein to the temptation of a good story. He most often presented him in the role of the man who planned his own funeral, the Jove who brooked no opposition, who always considered that he knew what was best for everyone, who demanded servility from his assistants and sometimes so intimidated his students that on social occasions they stood up when he addressed them—a vivid contrast to the admired, independent, unconventional, and rebellious Hilbert. When, however, I first read Courant's long article on Klein, published in *Naturwissenschaften* in September 1925, I was reminded of a remark he had once made to me about Hilbert's memoir of Minkowski.

"In that," Courant had said, "Hilbert revealed more of his own soul than he ever did at any other time."

In the *Naturwissenschaften* article Courant vividly recalled Klein's life. The gifted boy rebelling against the one-sidedness of the education offered by the humanistic gymnasium of his youth. The precocious university student, seventeen years old and already the assistant of Julius Plücker at Bonn. The recently promoted young doctor falling immediately under the spell of Göttingen. The independent scholar, stubbornly maintaining his independence from any "school," even Göttingen. The geometrically-minded and intuitive mathematician dramatically confronting in Berlin the abstract and arithmetical trend of the day. The fiery patriot, breaking off studies in Paris in 1870 to rush back to Germany to enlist in the army. The 23-year-old professor, rejected for Alfred Clebsch's chair in Göttingen because he was "too dangerous," delivering the famous Erlangen Program as his inaugural address at that university. The busy administrator and teacher of the Munich days, coming in contact for the first time with the applications of mathematics. The still young professor in Leipzig, at the height of his creative powers, locked in a furious race with Poincaré toward a theory of automorphic functions. The breakdown at the age of thirty-three which effectively ended Klein's career as a productive mathematician. The offer from America to succeed Sylvester at Johns Hopkins and then the offer from Göttingen. "The wonderful turning point"—the seemingly broken man who had lived another forty-three years and displayed the most varied sides as researcher, teacher, organizer, and administrator.

"What had been the secret of this personality and its workings?" Courant asked. "He possessed great power over men, because he combined spiritual superiority with accompanying objectivity; because he was not doing something for himself alone but was always directed toward his goal; because in the majestic dignity of his being there was no trace of vanity and no overweening opinion of himself. He did not lack the true humor which is the mark of real spiritual freedom. But all these things were eclipsed by the magic of his being, the magnetic power with which he was able to produce followers and to make co-workers out of even the reluctant."

Klein had lived long enough to see Göttingen once more an international center of mathematics and physics, foreign visitors coming again from east and west, the *Lesezimmer*—replenished and renewed—again the finest library of its kind in the world, an impressive ring of technical institutes—the result of a long collaboration between science and industry—surrounding the university and flourishing again. But he had died with the possibility of an institute for mathematics, which was to have been the climax of the whole development, still to all appearances hopelessly remote.

Then, just a few months after Klein's death, the city of Göttingen proposed a plan to build a new secondary school and to vacate the old school building, which was close to the physics institute and also to many of the other technical institutes. Such physical as well as intellectual proximity to the neighboring sciences had been an integral part of Klein's dream. Courant immediately tried to seize the opportunity, not only to bring together the various mathematical activities in research and instruction, scattered about the city, but also to relieve "catastrophic crowding" in Franck's physics institute. The government was willing to buy the old school building from the city and turn it into a mathematics institute. The only difficulty was—as Courant expressed it—"the city of Göttingen can give up its old school only if it can first build a new one and it can build a new one only if it first has the money—and it has no money."

In the late fall of 1925, during a visit with the Bohr brothers, Courant brought up this problem. The city needed 800,000 marks to build a new school, and such a loan was simply not obtainable in Germany at that time. Niels Bohr suggested approaching the International Education Board, which had recently made a grant to improve Bohr's institute in Copenhagen. Apparently Courant had never thought of the Board as a source of possible funds for the project, although he had already recommended several mathematicians for fellowships and had also asked for assistance in the purchase of foreign books and periodicals for German libraries. He immediately had a "fantastic" idea. Perhaps the Rockefeller people would be willing to lend the 800,000 marks to the city of Göttingen for the new school; then the university, with the money already promised by the government, would be able to purchase the old school. Niels Bohr advised him to forget the idea of buying and remodeling the old building and instead to apply for an outright grant. This, with the money available from the government, would make possible the erection of a new building specifically designed to meet the needs of the mathematicians and physicists.

Before the three friends separated, they decided that Courant should write a description of the problem and of his proposal in the form of a personal letter to Harald Bohr, who would shortly be in Paris, where the International Education Board had its European headquarters. Harald Bohr could then either discuss the idea with Augustus Trowbridge, a former Princeton physics professor who was the director for natural sciences, or simply present Trowbridge with Courant's letter, if that seemed more appropriate.

Trowbridge had visited briefly in Göttingen the previous fall to acquaint scientists there with the objectives of the fellowship program. It was, he had explained, less the desire of the International Education Board to improve the average of research output than to facilitate the development of exceptional scientists—or, as the French mathematician Émile Picard put it, "to make the high places higher rather than to fill in the valley with peaks." Trowbridge had also told Courant and Franck that they should consider what the International Education Board could do to help in the furthering of scientific life in Göttingen. With such an invitation, Courant did not feel that he was being presumptious in approaching Trowbridge, but still the sum involved was staggering.

"Do you think that we dare to speak to Mr. Trowbridge . . . of such an extensive request and that we are actually justified in thinking in these terms?" he wrote to Harald Bohr in Paris on December 22, 1925. "I personally believe that through the carrying out of this plan a completely unique center of mathematics and physics would be created here in Göttingen and that the Rockefeller people do not have to worry about wasting their money. But then I am not impartial."

Harald Bohr's response to Courant's letter and his report on his hour-long interview with Trowbridge could scarcely have been more encouraging:

"The chief thing is that Mr. Trowbridge has received your general proposal with great interest and warmth, and he is willing, already now—without any official steps having been taken—to discuss and pursue the idea."

By the beginning of 1926 Courant was in correspondence with Trowbridge and, at the American's suggestion, was sounding out the German government about its willingness to cooperate by contributing funds of its own to the project. Plans were soon being made for Trowbridge and George Birkhoff to come to Göttingen in the summer for further discussions.

Birkhoff, whom Courant had never met, was by then recognized both at home and abroad as the leading American mathematician. He was a professor at Harvard, the most recent president of the American Mathematical Society, and the youngest man elected up to that time to membership in the National Academy of Sciences. His role in relation to the International Education Board—as Trowbridge described it to Courant—was that of a traveling American professor retained to advise the Board in the field of his expertise.

Unfortunately Birkhoff's visit to Göttingen was to coincide with that of another, younger American mathematician who felt that he had many grievances against Birkhoff. It was thus to play a part in the development of an unforgiving enmity toward Courant on the part of Norbert Wiener.

Wiener had studied briefly in Göttingen before the war. On his return to America, in the course of delivering lectures at Harvard, he had run "into a series of logical difficulties which were clearly pointed out to me by Professor G. D. Birkhoff. . . .

"He was, as I was later to learn, intolerant of possible rivals, and even more intolerant of possible Jewish rivals. He considered that the supposed early maturity of the Jews gave them an unfair advantage at the state at which young mathematicians were looking for jobs, and he further considered that this advantage was particularly unfair, as he believed that the Jews lacked staying power. At the beginning I was too unimportant a youngster to attract much of his attention; but later on, as I developed more strength and achievement, I became his special antipathy, both as a Jew and, ultimately, as a possible rival."

Wiener had also spent the summer semester of 1925 in Göttingen and had received the impression that he had been encouraged by Courant ("an industrious, active little man who was eager to keep all the strings of mathematical administration in his hands") to apply for a Guggenheim Fellowship and return to the university the following summer. Courant had, Wiener thought, promised him "the full cooperation of my Göttingen colleagues in making my trip agreeable and in providing me with an assistant to help organize my papers and to take care of my lapses in German."

As it turned out, although Wiener was in Courant's opinion an "extraordinarily talented and powerful" mathematician, his presence in Göttingen in the summer of 1926 was an embarrassment. Wiener's father, a professor of languages who had pushed his son into college while the boy was still in knee pants, had been the most violent anti-German agitator in American academia during the war. The publicity which the young Wiener received at the time of the announcement of his fellowship ("I had been," he conceded, "a bit loquacious.") called the attention of people in the government to the fact that the son of this abhorred man was coming to study in Germany and was actually boasting about being welcomed there. Courant was requested by the *Kurator*—the representative of the government at the university—please, to try to keep Wiener inconspicuous. When Wiener arrived, recently married but alone, he found his welcome much less enthusiastic than he had expected. Courant "scolded" him for his newspaper publicity and did not give him the help and official recognition which he thought he had been promised. As a result, Wiener felt, his lectures were less successful than he wished, both as examples of mathematical research and as lectures in the German language.

Wiener, according to Friedrichs, was regarded in Göttingen as "rather uncultivated" in his mathematical writing and lecturing.

"Now perhaps 'uncultivated' is not the right word," Friedrichs said. "But I have studied some of Wiener's writing and one of his books I've read forwards and backwards. It *is* written in a clumsy way, not only in language, but also in substance. The mathematical argument is unusually inventive, original, surprising, but it is not streamlined. So to say, the judgment that it was 'uncultivated'—'clumsy' would have been the better word—was not quite unjustified; but the conclusion that the substance could not then be so hot, was wrong. And there we made a mistake in Göttingen."

Courant was disposed, he told me, "to do something for Wiener" because the American was a cousin of Leon Lichtenstein, a good friend. But it was only with great difficulty that he was able to persuade about twenty students to attend Wiener's lectures. As time went on, that number dwindled so embarrassingly that at one point he had to pay a student to attend.

Wiener had expected that the recognition he would receive in Göttingen would enable him to get out from under what he saw as "the continuous hostile pressure" of Birkhoff in the United States—and now Birkhoff was coming to Göttingen!

"[Birkhoff] represented the American whose support Courant most wanted. Courant approached me as an avenue through which he might win Birkhoff's goodwill. I told him that I had no influence whatever with Birkhoff and that Birkhoff's entire reaction to me was hostile."

By the time Wiener's bride arrived, Wiener felt that he was on the edge of a nervous breakdown. At this point his parents also descended upon the shaky new household, "partly to share in my supposed success and partly to keep a supervising eye on the newly married couple." The elder Wiener wanted to give a public lecture in Göttingen, and the son was forced to explain the rebuffs he had received. He found his father more concerned about the rebuff to himself. It was impossible to keep him from writing a letter to the ministry denouncing Courant.

The unhappy experience of Wiener was in complete contrast to that of another foreign visitor to Göttingen that same summer.

Since 1923 Alexandroff had returned each year, either alone or accompanied by countrymen. From 1926 through 1930 Courant always arranged for him to give regular lectures. Once the Russian mathematician taught three courses in topology, each for a quite different audience of mathematicians. The summer that Courant was trying to keep up some semblance of attendance at Wiener's lectures, Alexandroff's were crowded.

"Alexandroff and the other Russian vistors were very important, very influential," I was told by Herbert Busemann, who came to Göttingen in 1925 to study mathematics after "wasting"—as he said—several years of his life in business to please his father, one of the directors of Krupp. "The Russians filled a gap because they were familiar with certain more abstract tendencies which were not well represented in Göttingen. Courant, as probably many have told you, was rather reactionary in his mathematical outlook. He didn't see the importance of many of the modern things."

The subject matter of Alexandroff's work, unlike that of Wiener, was far removed from Courant's own—very pure and very abstract. Courant, however, was extremely enthusiastic about it, as he was always to be about the work of the many other Russians whom Alexandroff introduced into Göttingen over the years. They were all "of the first rank," he wrote somewhat later. The blind Pontryagin, who had always to be accompanied, was "an absolutely leading spirit in topology." Gelfond "had astonished the world" with his proof of the transcendence of certain famous numbers; the work on prime numbers by Schnirelman, who walked barefoot through the streets of Göttingen, "likewise." Lusternik "had the most highly original thoughts" concerning topology and analysis. Kolmogoroff was "an absolute master" in a great variety of fields.

To Alexandroff this enthusiasm for the work and the success of other mathematicians was impressive.

"There is very seldom the combination of really great scientist and great human personality," he said to me when I talked to him in Moscow in 1971. "And the emotional nature of Courant, his unselfish relation to other men, his interest in other human beings, his lack of egotism in relation to the world—he is absolutely exceptional. I do not know of any other man who has this kind of personality, this unselfish position in the science, and this *joy*—do you understand me?—in the success of other, younger men around him. Courant was himself quite a young man at that time, but he was like that even then; and I think that really explains him and his success."

The Göttingen summer of 1926 had a special significance for Alexandroff, because it was then that he met Heinz Hopf, with whom he soon established a professional collaboration and a friendship that was to last until the end of Hopf's life. Although Hopf was two years older than Alexandroff, he had only recently taken his doctor's degree in Berlin. This lateness was partly because he had served on active duty as an artillery officer during the war and partly because he had developed very slowly as a mathematician. Courant was also enthusiastic about Hopf. Recommending him the following year for a professorship in Switzerland, he predicted "in a few years he is sure to be known throughout the world."

Even in those long ago days, Alexandroff wore extremely thick lenses and was already quite bald. When I visited him in Moscow, he still looked very much as he had in snapshots I had seen of him from the 1920's. On a table in his crowded, rather Germanic living room there was a photograph of Urysohn, who, swimming with Alexandroff on the coast of Brittany in 1924, had been dashed against the rocks and killed at the age of twenty-six. I had recently left Courant, tired and depressed, in America. Hopf had just died in Zurich. Alexandroff was recuperating from an almost fatal illness which had prevented him from attending Hopf's funeral. We talked for quite a time about that long ago summer of 1926, when he had first met Hopf, and the other summers which he had spent in Göttingen.

"It was a beautiful time in my life," he told me as I prepared to leave, "and I cannot forget it yet."

That summer of 1926 was also the summer when Birkhoff came to Göttingen to discuss the proposed grant to the university from the International Education Board. The unhappy and lonely Wiener tactfully kept out of the way while Courant and Franck met with Birkhoff and Trowbridge for two days.

The most extravagant and completely satisfactory plan considered at the meeting was the erection of a new building for mathematics and the improvement of the already existing physics building. Not quite so satisfactory but still very nice was the purchase and remodeling of the vacated high school building and the including in it of the necessary additional facilities for physics. If neither of these proposals was acceptable, Courant still had hopes that the International Education Board would grant funds for some improvements in the existing mathematics and physics facilities.

No final decision was reached during Birkhoff's visit, but it was agreed that detailed sketches and estimates of cost would be made for all three proposals. For this work Courant turned to Neugebauer, whose artistic gifts and knowledge of engineering, draughtsmanship, and the needs of the mathematicians made him the perfect choice. Courant and Neugebauer were eleven years apart in age and of completely different characters. "Courant's word was *maybe*," Hans Lewy said to me. "Neugebauer's was *yes yes, no no*." But the two were fast becoming an administrative team, the purposeful disorganization of Courant balanced by the efficiency of Neugebauer. By the end of October 1926, working together, they had managed to get all the necessary plans and estimates to the International Education Board.

At Christmas word came at last from Trowbridge.

"I may say informally . . . ," he wrote to Courant, "that the action of the Board was to vote into the hands of the Executive Officers of the Board a sum not to exceed $350,000 for the construction and equipment of a building for the Mathematical Institute, and for the construction and equipment of an addition to the Building for the Physics Institute of the University of Göttingen, with the understanding that the Prussian government will provide annually a sum not less than $25,000 to cover additional maintenance costs of the Institutes of Mathematics and Physics."

Trowbridge's letter was dated December 21, 1926, a year and a half after the death of Felix Klein.

"So Klein never knew?" I asked Courant as we sat in his office on the thirteenth floor of the Courant Institute of Mathematical Sciences in New York City.

He gazed out the large corner window at the wooden water towers topping the buildings of Greenwich Village.

"No," he said. "Klein never knew."

TWELVE

THE YEARS during which a mathematics institute was coming into existence in Göttingen were to have about them, in memory, a kind of glow.

In comparison to the hard post-war period, the country was enjoying a rather remarkable prosperity. The old general, the victor of the Battle of Tannenberg, was lending respectability to the beleaguered new republic. Foreign relations were improving. In 1926 Germany was admitted to the League of Nations.

In Göttingen there was again a scientific paradise. Gifted students flocked to the university. There was a constant procession of distinguished visitors from all over the world. Sometimes they came merely for a single talk or a series of talks to the *mathematische Gesellschaft*. Often they lectured for a full term as guest professors. The very air seemed to Courant to crackle, as it had in his youth, with scientific electricity.

Courant himself, in his late thirties, could view his situation with satisfaction. He was a German professor, a uniquely comfortable, secure, and respected person in the Germany of the time and in a small university town like Göttingen. His personal life was highly satisfactory. Two maids relieved Nina of the responsibility of house and children. She concentrated on her own playing of the viola da gamba, an old instrument of the viol family, and also directed a group that specialized in the singing of choral music. There were yearly skiing trips with family and assistants to Arosa in Switzerland.

Courant's relations with the ministry were as close and friendly as Klein's had been. His opinion was sought when university appointments were made. He had considerable influence with the International Education Board. His files are full of letters beginning "I am writing for Kurt Hahn, a friend of our Göttingen circle," "for my old teacher, Professor Adolf Kneser," "for my colleague Franck." Trowbridge relied upon him to evaluate other professors' recommendations for Rockefeller Fellowships, "since you are yourself so familiar with the type of man we are trying to help with these fellowships."

In scientific publishing Courant and Springer were an influential team. The *Mathematische Zeitschrift* was followed by a comparable publication in physics. The old established journals and the mathematical encyclopedia had fallen years behind the scientific work actually being done. Now, for a change, important work in both mathematics and physics got into print as soon as it was completed, fields where current interest lay were promptly explored by competent authorities. The Yellow Series, emphasizing the connections between mathematics and physics, grew volume by volume.

Although Carl Still was, as Courant always said, "no Krupp," he stood generously by with financial help where it was needed. His friendship with Courant extended to Courant's friends and colleagues. On the Still estate they hunted hare, discussed their scientific problems, and played with scientific toys. Franck and Pohl set up a laboratory. Runge installed a telescope in an observation tower. Prandtl investigated the irrigation problems of the estate.

In 1925, with Friedrichs as his assistant, Courant began to work on a second volume of Courant-Hilbert. He also decided to edit and publish his calculus lectures.

Although there were a number of calculus books in print, he thought it would be very difficult for the beginner to locate a book which would open up to him "the direct way into the living essence of the subject" and give him "freedom of movement in relation to the applications." In his book, as in his lectures, he wanted to show the close connection between analysis and applications and to emphasize intuition as the original source of mathematical truth. Most important, he intended to proceed as directly as possible to interesting and fruitful topics.

The calculus was the first of Courant's books published under his name alone. It might well have been called the Courant-Klein; for, as Courant acknowledged in his foreword, "What I have here attempted is completely in the direction of my great predecessor, Felix Klein."

In spite of the single authorship, the calculus, which was published in two volumes in 1927 and 1929, was a communal effort.

Notes on Courant's lectures, written up by earlier assistants, were already in existence. These were sent to Springer and put into type. Then the galley proofs arrived. The rooms in which the *Praktikum* took place became the scene of "proofreading festivals." All the assistants took part, all offering—as Neugebauer was later to recall—"simultaneously uttered and often widely diverging individual opinions about style, formulations, figures, and many other details."

The corrected galley proofs then went to Courant and then back to Springer. The changes were made. Fresh galley proofs arrived in Göttingen.

Friedrichs shook his head, remembering.

"And they were again full of mistakes, because as in the beginning

Courant did not bother about technical details. So we assistants corrected these mistakes, and then we had first page proofs and even second page proofs. The outcome—well, you could not recognize it from what had been there originally."

The mystery was that the final product of this process was unmistakable and pure Courant.

"People who have not participated sometimes say that Courant's books were written by his assistants," Friedrichs said. "But this was not true. In a way Courant *would have liked* his assistants to write them. But they couldn't do it. Essentially the books were always Courant's books."

The assistants who supervised the *Praktikum* were talent scouts as well as proofreaders and editors. At the beginning of the term in 1925, they were delighted to discover that the answers of a new student from Yugoslavia were invariably correct and that there was no longer any need for them to solve the problems themselves. They promptly alerted Courant to the presence of Willy Feller. After the third calculus lecture, to Feller's amazement, the professor—an unbelievably august personage to a European student of that day—approached. Questioning the boy about his education in his native land, Courant discovered that Feller was already doing mathematics on his own. He told him to bring his work to the next lecture. Even thus instructed, Feller was too bashful to produce his papers on the appointed day. The next morning he was awakened by a commotion on the stairs leading to his attic room. There was a knock on the door. Courant entered and left a few moments later with the desired papers.

After Feller was "discovered" in the *Praktikum,* he was an accepted member of the new "in group" which gathered around Courant. As in the past, there was no clearly defined standard of admission to this group. The year after Feller came, Franz Rellich turned up in the *Praktikum.* He was an Austrian, only twenty years old but of such charm, open-mindness, and independent judgment that he was accepted into the group even before he had proved himself mathematically.

One other young man who came to Göttingen during this time also should be mentioned; for although he was never a member of Courant's group, Courant played a role in his coming.

At the beginning of 1926 Hilbert proposed to the International Education Board that a 22-year-old Hungarian, Janos (John) von Neumann, be given a fellowship to come to Göttingen and work with him. At this time von Neumann did not yet have his Ph.D. Trowbridge replied that, although the degree was not a *sine qua non* for a fellowship, "in this case I rather judge that, combined with [the] youth of the candidate, he would probably not fall into the class which we are trying to reach with these fellowships—that is, the class of a man well advanced in his career, who has already published a sufficient amount so that one can form a judgment of his technical ability."

While the ailing Hilbert was in Switzerland, Courant took up von Neumann's case.

"It appears to me as though Hilbert somewhat unclearly expressed himself in his first letter," he explained to Trowbridge. "Mr. von Neumann, in spite of his youth, is a completely exceptional personality . . . who has already done very productive work . . . and whose future development is being watched with great expectation in many places."

The work for which Hilbert wanted von Neumann was of the greatest significance in Courant's opinion—"it concerns investigations in the foundations of mathematics and a program on completely epoch-making lines initiated by Hilbert . . . which Mr. von N. has already pursued independently."

There was also another consideration. Hilbert had failed greatly since Trowbridge had last seen him. Although it was hoped that as a result of a treatment for pernicious anemia recently discovered in America he would be able to overcome the most serious aspects of his illness, there was no time to be lost in promoting the carrying through of his scientific program: "I believe that in this way one really can perform an important service to science."

Under the force of Courant's plea, Trowbridge reconsidered Hilbert's request; and in the fall of 1926 von Neumann came to Göttingen as a Rockefeller Fellow. The young mathematicians there recognized that he was obviously a prodigy, but some were suspicious of what they saw as a certain "glibness" about him. They also found his mathematics "too abstract" for their taste.

"We were wrong about that," confessed Friedrichs, part of whose later work was to be strongly influenced by the work of von Neumann.

Friedrichs told me that he and Lewy were also somewhat skeptical of the developments which were taking place in physics.

In 1925 one of Born's assistants, Werner Heisenberg, formerly a student of Sommerfeld in Munich, had created quantum mechanics. Born and Pascual Jordan (who had come to Born from Courant) had followed this with a striking mathematical formulation which rounded out the discovery and had published this work with Heisenberg.

"I was very much repelled by what the physicists were doing," Lewy admitted to me. "To my mathematical mind they were too sloppy and also their way of talking was so glib that I had the impression—which was wrong, of course, as it turned out—that they were fourflushers. Often if you would corner them, ask them to explain precisely, they would be evasive or else you would find out that they didn't quite understand what they were saying themselves. They obviously had some physical intuition which I didn't have, but their mathematics was objectionable. The type of personality that was needed at that time in physics was not to my liking. Well, what happened teaches one humility, and it shows that different situations in science require different types of personalities. So one cannot make up one's mind what is *the right way* of being for a scientist."

I asked Courant if he had had any such qualms, and he said no. Within a few weeks Niels Bohr had publicly placed his stamp of approval on Heisenberg's work.

The physics being done at this time in Göttingen and other places was to result in a dramatic and unexpected professional triumph for Courant.

Heisenberg's quantum mechanics of 1925 was followed by Erwin Schrödinger's apparently different wave mechanics in 1926. Physicists struggled to deal mathematically with strange and unfamiliar ideas.

In Schrödinger's theory (which Schrödinger himself soon showed to be mathematically equivalent to Heisenberg's theory), a stationary state of a physical object, such as a hydrogen atom, is described by the solution of a certain partial differential equation which has solutions only if the value of a certain parameter, representing the energy of the object, is one of a particular sequence of numbers. When the numbers in the sequence were recognized as the eigenvalues of the equation and the stationary state was seen to be described by the associated eigenfunction, the physicists turned to the mathematicians to find out where they could learn about eigenvalues and eigenfunctions. To their surprise they were told that exactly what they wanted was ready and waiting for them in Courant-Hilbert, the book which they had dismissed several years earlier as nice but somewhat old-fashioned—obviously irrelevant to the modern physics of the quantum.

The particular importance of Courant-Hilbert for the physicists was not, however, the new mathematical material it contained, such as Courant's own theory of eigenvalues and eigenfunctions. Rather it was the exposition of the classical theory, which Courant had laid out in detail as the foundation for the presentation of his work.

The story of Courant-Hilbert had been told many times by historians of science.

"The book, in compact, convenient form, contained practically every mathematical method, trick, device, and special detail required for the development of the Schrödinger theory, not to mention much that was applicable to the theory of Heisenberg," Banesh Hoffman wrote in his book on the story of the quantum.

"In retrospect, it seems almost uncanny how mathematics . . . prepared itself for its future service to quantum mechanics," Max Jammer observed. "Published at the end of 1924, [*Methoden der Mathematischen Physik*] contained precisely those parts of algebra and analysis on which the later development of quantum mechanics had to be based: its merits for the subsequent rapid growth of our theory can hardly be exaggerated."

It was, E. T. Bell said simply, one of the most dramatic anticipations in the history of mathematics.

As it happened, during the same period that physicists were finding a need for the mathematical methods of Courant-Hilbert, Courant was engaged in another work which would also turn out to be a dramatic anticipation.

Since his student days, when the much admired Walther Ritz had utilized Hilbert's approach to Dirichlet's principle to develop an efficient numerical method for solving the boundary-value problems of partial differential equations, Courant had been intrigued by the fact that the same methods which lead to proofs of existence, as in the case of Dirichlet's principle, can also be useful in that most practical chore of mathematics—numerical calculation. Was it possible, he now wondered, to proceed in the other direction? Could a method of approximation used for numerical calculation also be employed for existence proofs?

The particular method which Courant proposed to investigate was one long used by engineers to obtain *approximate* solutions of problems the solutions of which are given *precisely* only by the solution of the corresponding partial differential equation. In this method of "finite differences," the plane is conceived as a lattice of a given mesh. The values of the function are sought only at lattice points rather than at every point of the plane. A suitable finite difference equation can then be taken as a substitute for the harder to handle, partial differential equation.

It was Courant's idea to investigate the purely mathematical significance of this method of the engineers—to take it with full mathematical seriousness. First, he suggested, it should be determined whether finite difference equations are meaningful in a mathematical sense; that is, whether solutions of them in fact exist. Second, it should be determined whether, when the mesh is taken sufficiently fine, the solutions of the finite difference equations actually do approximate the solutions of the corresponding partial differential equations.

In 1925, in his first work on these questions, Courant carried out the finite difference method for the same partial differential equation problem to which the Dirichlet principle was applied. One of his main tools was the lemma which he had employed in earlier work on Dirichlet's principle.

The partial differential equations which Courant treated were of the so-called elliptic type. Such equations govern equilibrium; for their solution conditions have to be prescribed on the boundary of a body. Equations governing motion are hyperbolic; the heat equation, parabolic. After Courant's work the question naturally arose whether the method of finite differences could be applied to hyperbolic and parabolic equations. For the solution of these equations, conditions are prescribed at an initial time.

"It is a remarkable fact," Friedrichs told me, "that on purely mathematical grounds elliptic equations would not have acceptable solutions if initial conditions were prescribed for them; and hyperbolic and parabolic equations would not have solutions under boundary conditions. This fact—which the French mathematician Jacques Hadamard pointed out in the introduction to a very long book in 1921—was emphasized over and over again to us by Courant. At first it may seem trivial. It is not really obvious that it is deep. But it is. Very deep. I think it is one of Courant's definite merits that he spread the news, as it were, about this idea of Hadamard's. It has had a great effect in mathematical analysis, but I think it might have gone unnoticed if it had not been for Courant."

As always, following the admired example of Hilbert, Courant proceeded to bring his new work before his students. In 1926 he conducted a seminar on finite difference equations and suggested to Friedrichs and

Lewy that they should see what they could do in the hyperbolic case. This was the summer of Norbert Wiener's unhappy experience in Göttingen. He attended the seminar; but, as Friedrichs remembers, he did not give a talk. He seemed, however, quite interested in the work reported on at several seminar meetings.

That fall Courant had some more results on the elliptic equations for the equilibrium problem. Friedrichs and Lewy had some results on the hyperbolic equations for wave propagation. Courant suggested that the three of them present their work in a common paper.

In future years the Courant-Friedrichs-Lewy paper, published in 1928 in the *Mathematische Annalen*, was to play a considerable role in numerical analysis. In particular, certain severe restrictions—the need for which was first recognized by Lewy—were to become of fundamental practical importance with the development of computing methods during the Second World War.

While working on the finite difference approach, in another joint work, Friedrichs and Lewy had also been trying to approach directly the related hyperbolic partial differential equations. They had recognized that they could establish the uniqueness of the solutions of these by a certain "energy inequality," but they were baffled by the problem of establishing the existence of the solutions. Then one day when they were discussing their work on a walk in Göttingen—Friedrichs still remembers the spot exactly—Lewy suddenly stopped. "I've got it! I've got it!" he shouted. "We can use the energy inequality for existence, too!"

"I was absolutely dumbfounded," Friedrichs recalled. "It was an uncanny insight. A considerable part of the existence theory of partial differential equations has since stemmed from this idea. I think it is a good example of Courant's influence as a teacher that it was his interest in the subject that drew Lewy into it."

Once I asked Courant who of all his students had most *surprised* him, meaning which one had become a greater mathematician than he had expected him to become. At first he misunderstood my question and thought I was asking which of his students had had the greatest ability to surprise him. To this question, his answer was "Lewy—he is a most original mathematician." When I repeated my question, however—understanding this time—he replied without hesitation, "Friedrichs."

"I do not think, however, that most teachers would have perceived Friedrichs's ability as early as Courant did," Lewy told me.

"But you said how greatly you were impressed by Friedrichs when you first came to Göttingen," I reminded him.

"Yes, but I was a fellow student. I think it took a keen observation as well as a really intense human interest on Courant's part to see what was there. Friedrichs was very shy. Courant forced him out into what were difficult situations for him. He sent him to work with von Kármán at the aerodynamics institute in Aachen. I went to the station with Friedrichs. He was a very unhappy young man. If it had not been for Courant, he might have just pulled into himself and become a gymnasium teacher."

The proof sheets of the final common work with Courant were shown by Friedrichs and Lewy to van der Waerden, who was an active member of the Göttingen circle at that time. When I interviewed him in the summer of 1971 in Zurich, he still spoke with enthusiasm—almost half a century later—of the vivid impression made on him by the paper on the *existence* and *uniqueness* of the solutions of partial differential equations in the elliptic, hyperbolic, and parabolic cases. Before he came to Göttingen, he told me, he had agreed to tutor an engineer who wanted to learn something about partial differential equations so that he could solve an equation relating to the conduction of heat in a cylinder.

"I had a bad conscience about him because I had taken his money and he hadn't learned anything useful. So when I came to Göttingen, my mind was much more open to these important things. All my life," he added, "I have had great advantage from having read that paper at that moment."

At the time of publication of the Courant-Friedrichs-Lewy paper, however, engineers did not pay any more attention to the mathematicians' results on finite difference equations than physicists had paid to the mathematical methods of Courant-Hilbert when it first appeared.

Courant did not consider himself an applied mathematician, but he always believed that mathematics includes the applications and cannot be separated from them. When Runge died in 1927, Courant expressed this feeling in an article about his father-in-law—which had unexpected repercussions.

In the article he pointed out that when, in 1904, Klein had established Runge in Göttingen as the first professor of applied mathematics at a

German university, Klein had taken "the decisive step to retrieve for the applications their proper place in our science." That Klein, "who always strove to preserve science as a unified whole," had felt it necessary to take such a divisive step had been due to the conditions in mathematics at the time. But, in Courant's opinion, such a separation between mathematics and applied mathematics was no longer necessary. It was Runge's great achievement that he had repaired the broken connections with the applications and restored the unity of mathematical science, "which *includes* the applications." Thanks to Runge, it was no longer necessary to designate someone as "a professor of applied mathematics."

Such heresy brought a strong rebuke from Richard von Mises, professor of applied mathematics at Berlin.

"It is a monstrous misconception of the present situation to think that one can now renounce the cultivation of applied mathematics in special institutes and with special teaching positions."

Courant held firmly to his position.

"I would go a step further than Mr. von Mises, in that I consider we should strive for the desirable goal that *no* teacher of our subject stands coolly apart from the applications. . . . But where I am absolutely in opposition to Mr. von Mises is in the idea that it is possible to strengthen the applications by separating them from the purely theoretical science."

It was during this same year (1927) that Trowbridge asked Courant to compile a "Who Is Who" in American mathematics for his own use and that of the International Education Board.

Courant didn't feel that he was really qualified to make such an evaluation, but he responded by listing American names with which he was familiar. He also added some general observations about contemporary American mathematics.

Great improvement had taken place in the last ten or twenty years. A truly native scientific culture had developed. And it was impossible for anyone to predict what might come out of it in the near future. The predominant characteristic of American mathematicians seemed to be a tendency to favor the abstract and the so-called pure areas of mathematics. Their greatest success had been in topology. Princeton, especially, and Harvard were without a doubt the best places in the world to study that subject. Applied mathematics, however, was treated like a stepchild in America. There was no real contact between mathematics and physics— as far as Courant could see—and hardly any between mathematics and technology.

"But," he told Trowbridge, "that situation too could quickly change."

For Courant post-war Göttingen was always to remain in memory a paradise where the distinction between "applied" and "pure" mathematics did not exist, a place where there were "mathematicians, abstract mathematicians and *more concrete* mathematicians and physicists talking to each other quite intensely and very frequently—and understanding each other." In the years to come, when he was no longer in Göttingen, he was to see it always as his task "to restore this easy communication" between mathematics and its applications and "to bring mathematics back into the mainstream of science."

THIRTEEN

IN SPITE OF the glow which surrounded the Göttingen of the 1920's, it should be said that there also existed in those years a certain amount of hostility toward the scientific faculty of the university both in Germany and abroad.

Many on the outside deeply resented the tendency of the people in Göttingen to consider themselves at the center of the world and to be careless in paying attention to work done by others at other places. In 1928, when the Courant-Friedrichs-Lewy paper appeared in the *Mathematische Annalen*, Norbert Wiener pointed out that in Courant's part of the paper he had introduced, among other ideas, the probability interpretation of finite difference equations without mentioning a similar interpretation which had earlier appeared in published work by Wiener and Phillips.

Friedrichs shook his head ruefully as he said to me, "*It was always happening to him.* Norbert Wiener was very offended. I have never really understood the justification. Courant didn't know about Wiener's work. I don't think Wiener mentioned it in Göttingen—although he may have. Maybe he mentioned it to Courant, and Courant forgot that he had mentioned it to him and thought it was his own idea. These things happen whenever people work together in the same field. It is rare in mathematics, but in fields like physics it happens all the time. Anyway, later, Courant was always very careful to mention Wiener and also several other authors of earlier relevant work which he hadn't known about at the time."

Much later Wiener, still unforgiving, circulated in manuscript a novel in which the character of a professor, who was unkind to gifted young people but took over their ideas, was rather obviously based on Courant.

"It was a common failing of Göttingen people that they were not very conscientious in attributing," Lewy conceded to me. "That was true for almost all of them—Hilbert included. But when you look back at papers by some of the great heroes of mathematics, you very often find that they are careless in these matters. That greater care is taken now is, I think, due to the fact that jobs depend to a more explicit degree on the credit that a person is given. It is undoubtedly true that the group in Göttingen was careless about studying what other people had done and attributing their results to them, but I think this must be seen against a background of less care."

The Göttingers had a facetious expression for the process of making someone else's idea one's own. They called it "nostrification." There were many levels of the process: "conscious nostrification"—"unconscious nostrification"—even "self-nostrification." This last occurred when one came up with a marvelous new idea which he later discovered had already appeared in earlier work of his own.

In addition to some general hostility toward the mathematicians of Göttingen, there also existed in Germany by this time a certain amount of what Courant described as "unfriendliness" toward him personally. Most of his critics objected especially to what was accepted by his friends as "Courant's way" and found "more funny than offensive."

"Ja ja," Neugebauer agreed. "Courant's way of doing things and being active in a certain direction irritates many people. I know that well. And then there is his way of saying something and taking it back. His probing. That irritates people. And his indecision causes him trouble, I am sure. He wavers, and people think he has something secretly behind it. And yet the loyalty of his own group was always very great. He held his group together better than most people. There is no doubt about that. It is certainly a funny situation."

I remarked that someone had said to me that Courant was a person you could like and yet still be completely aware of his faults.

Neugebauer nodded.

"One takes that. So it is."

In this same connection Hans Lewy pointed out that Courant's irritatingly vague, almost inaudible manner of speaking might be a reflection of the kind of things that he did in mathematics.

"In some ways analysis is more like life than certain other parts of mathematics," Lewy explained. "For instance, than number theory. There is nothing in life like number theory. It is too clear-cut, too straightforward, too black and white. By that I am referring to classical number theory. When you come to analytic number theory, things change. Analytic number theory has the same shortcomings or, if you wish, the same advantages as analysis."

In Lewy's opinion the statements of analysis—and particularly that part of analysis which deals with partial differential equations—"are kind of hesitant statements."

"There are some fields in mathematics where the statements are very clear, the hypotheses are clear, the conclusions are clear. The drift of the subject is also quite clear. But not in analysis. To present analysis in this form does violence to the subject, in my opinion. The conditions are or should be considered temporary, also the conclusions, sort of temporary. As soon as you try to lay down exact conditions, you artificially restrict the subject. Therefore, I think in a sense Courant's way of talking is—well, you don't know—cause or effect of the subject he is dealing with."

"Courant is suspicious about clear-cut and definite formulations of the truth," Friedrichs contributed. "Well, he knows there are cases where such statements can be made. Still, he is doubtful. Before he is willing to nail himself down in a definite statement—he is very reluctant to do that, because he feels that the essence, or whatever you want to call it, is what counts more. He is inhibited out of a philosophical insight, so to say."

Besides Courant's irritating indecision, there were other things about him which were criticized. Mathematicians who were not his friends spoke of his tendency "to get himself counted," "to organize," "to meddle." Even his obtaining the grant from the International Education Board was looked upon in some circles as "undignified begging abroad."

In 1928, when the new institute was almost finished, Courant and the other Göttingen mathematicians found themselves in opposition to a number of German mathematicians, especially those in Berlin.

That spring L. E. J. Brouwer (although Dutch, very pro-German) had addressed an open letter to the members of the German mathematical society in regard to the international congress to be held late in the summer in Bologna. Repeating violent words uttered by the French mathematician Painlevé during the war—"and not yet retracted"—Brouwer had questioned whether it was possible for Germans to attend the congress "without mocking the memory of Gauss and Riemann, the humanistic character of mathematical science, and the independence of the human spirit."

Harald Bohr, in what he described as his role as "ambassador from the international world," came to Germany and visited a number of universities. He reported to Hardy on the "rather excited" situation—in his English, "which I remember you were kind enough to call a 'language' even if it was not ordinary English."

"The word *Germans* in connection with such a phrase as *whether the*

Germans will go to Bologna or not is more physically than mathematically expressed," Bohr wrote. "It is not dealing with realities, because there are just in these questions all sorts of Germans and even among the internationalistic Germans there are great differences in respect to what they call clearness concerning the international character of the congress. To take two opposite persons both completely agreeing with one another and with us in the principal view that politics and science must be completely separated from one another; namely, Hilbert and Erhard Schmidt [one of Hilbert's earliest students and a professor in Berlin]. The first thinks that the international character of the congress is clear enough and will go at any rate (if there will be no earthquake), while the other thinks that he cannot go if there will not be obtained much greater clearness."

Ludwig Bieberbach, also a professor at Berlin, followed Brouwer's letter with a letter addressed to the German academies and the rectors of all advanced educational institutions, urging that they prohibit their members from attending the congress. Hilbert responded with a strongly worded letter supporting attendance at the congress. Richard von Mises, professor of applied mathematics at Berlin, urged that "no representative of applied mathematics" go to Bologna. Other German mathematicians allied themselves on one side or the other for various, often individual reasons. Hardy joined the Italian mathematicians in their insistence that the congress was to be truly international and was not a front, as claimed by some, for the organization which had vengefully excluded the Germans from a congress held in 1924 in Toronto.

Like the other Göttingen professors, Courant accepted the assurances of the Italians as to the international character of the congress. He felt that it was impossible to boycott it "without appearing to compromise our willingness for reconciliation and our desire for peace." He regretted, however, the personal tone which had developed in the debate, especially in the exchange between Hilbert and Bieberbach. He wrote a long friendly letter to Bieberbach, whom he had known when they were both students in Göttingen, and attempted to explain Hilbert's motivation and the fact that his letter had not been directed personally against Bieberbach. Hilbert had already agreed to go to Bologna before Brouwer's letter had appeared, and he had "passionately rejected this interference and judgment from the outside."

"Why is Brouwer's interest in the subject interference," Bieberbach demanded tartly, "when that of Bohr and Hardy is not?"

In the end Hilbert led a group of seventy-six Germans to Bologna. But the battle over attending the congress and a subsequent smaller scale battle, in which Hilbert forced Brouwer out of his position on the editorial board of the *Annalen*, left a residue of bitterness in Germany against the mathematicians of Göttingen and their foreign friends.

It was not long after the congress in Bologna that the building on Bunsenstrasse was ready for occupancy. The thought of the "institute" filled Courant alternately with pride and embarrassment, satisfaction and dismay. As a German professor he had a very easy, pleasant life. As the director of a large and generously budgeted facility, he would have "to march always with a heavy knapsack on my shoulders." Forty years later he could still recall vividly how he had walked around the old wall of the city with a friend and debated his decision.

"And then this friend said this strange thing: 'If in a few years it turns out to be worthwhile, then you should do it. That would be wise. But if it turns out to be a disappointment, then it would be *such* a waste of energy!'"

Since I knew that less than four years after the dedication of the Göttingen institute Courant was removed from his position as its director, I started to ask—

"But it turned out to be very worthwhile," he said with satisfaction, even before I had framed my question. "Our institute in Göttingen—you have seen it?—it was unique in the beginning. Then gradually the idea spread, here and in Germany. Now there are wonderful modern university buildings for mathematics everywhere. It is no longer such a novelty. But it was the beginning, our institute in Göttingen."

The new building—a three-level T-shaped structure—provided everything the mathematicians had ever needed or wanted in their physical surroundings. Courant always gave Neugebauer the credit for its planning. The basement contained such requirements as a bicycle room, a book bindery, and a room for refreshments. The double stairs of the main entrance led to a spacious lobby, which today contains a bust of Hilbert and is known to students and faculty as "the Hilbert space." On the main floor there were two large auditoriums, *Maximum* and *Minimum*, and four other rooms of various sizes—all equipped for lectures in applications as well as in theory—a mechanical drawing room, a room for the meetings of the *Praktikum* and of the *mathematische Gesellschaft*, smaller meeting rooms, offices, individual workrooms, consulting rooms. On the top floor were the spacious and well-planned quarters of the *Lesezimmer*, still the heart of the mathematical life of Göttingen.

Among the visitors to the new building were Artin, by then a professor at Hamburg, and his young wife Natascha. I asked Natascha Artin, now Brunswick, for a "picture" of Courant at this time; but although she remembered Nina vividly—"she was a *very* striking woman"—she could not really tell me how Courant looked.

"What I remember is that he was always surrounded by a crowd of people wherever he went. Sometimes you wouldn't even see him. You would just see the crowd moving from one place to another."

The formal dedication of the Mathematics Institute of the University of Göttingen took place on December 2, 1929. The featured speakers were the long-ago rivals, Hermann Weyl and Theodor von Kármán. Courant, Born, and Franck had long wanted to bring both of them back to Göttingen. Up to that time, Weyl had chosen to remain in Zurich. The calling of von Kármán (proposed in 1925) had never been pursued. Much as they would like to have had Kármán with them in Göttingen, his friends were well aware that in the opinion of some professors there were already too many Jews on the scientific faculty.

Standing before his Göttingen colleagues and the great crowd of visitors in the auditorium of the handsome new building, Courant was reminded of the anguished feeling of a warmly clad man who walks through a poverty-stricken section of town on a cold winter day.

"Involuntarily one asks oneself why he is in such a position. . . ."

The new building had come about as the result of decades of work on the part of Felix Klein, he told his audience. It had been Klein's belief that a mathematician is not a self-sufficient creature who needs only paper and pencil to create. He must have as well a library, models and other demonstration materials, instruments; and he and his science must stand "in active and reciprocal relation" to the other sciences and to society. And yet, to justify the existence of the institute, it was not enough—in Courant's opinion—merely to cite the historical development and the interaction the institute would provide between mathematical and other interests.

"The ultimate justification of our institute rests in our belief in the indestructible vitality of mathematical scholarship. Everywhere there are signs to indicate that mathematics is on the threshold of a new break-through which may deepen its relationship with the other sciences and demand their mathematical penetration in a manner quite beyond our present understanding. It is not at all certain whether in the course of such a development the so-called applied sciences may not have to take second place to things now considered remote abstraction. . . ."

The old Hilbert was exultant.

"There will never be another institute like this! For to have another such institute, there would have to be another Courant—and there can never be another Courant!"

The dedication of the new institute took place a little more than a month after the American stock market crash. Signs of slump were already appearing in Germany, unemployment up, wages down, bankruptcies increasing daily. But the mathematicians, happily installed for the first time in a building of their own, were unaware of them.

With Hilbert's retirement in 1930, all the heroic figures of Courant's first student days were gone from the active faculty; but, as in the past, each of the mathematics professors of Göttingen had a highly individualized approach to his research and to his teaching which attracted a particular type of student.

Landau was the senior member of the new faculty. During the 1920's he had been the center of a fantastic flowering in the analytic theory of numbers, and he had written a number of influential books on that subject. In lectures, as in books, his ideal was absolute rigor and completeness. The assistant was instructed to interrupt if the professor omitted anything at all. Standing before the big blackboards in the lecture halls of the new building, Landau wrote with speed—theorem, proof, theorem, proof— while a menial with a sponge hastened after him to erase what he had written so that there would be room for him to write more. He never gave any explanation of where he was going, but he had organized his material so well that there was—I have been told—a quality of incredible clarity about his lectures.

Being Landau's assistant was a wearing experience. Werner Fenchel, who held the position for several years, recalls that when he arrived in Göttingen, Landau demanded to know when he got up in the morning. "Oh, about eight," Fenchel hazarded. The next morning, precisely at eight, Landau presented himself at the door of Fenchel's room. "But I learned to adjust," Fenchel smiled. "Landau took things very literally, always. In the beginning I would say, 'I have an appointment,' and he would say, 'That you can change.' But then I could tell him straight, 'I have no time.' That he would accept."

Unlike Landau, Herglotz ranged over a variety of areas, which included number theory, celestial mechanics, continuous groups, the mechanics of particles and continua, optics, to name a few. It was his custom, almost without exception, to offer a new course each semester.

"His lectures were works of art in every respect, including the lecturer himself," Charlotte John told me. "There was something of the nineteenth century about him, which he cultivated, I think. He had the elegance of that century. In fact, his dress was actually like that of Goethe. And he had really luminous eyes. He looked like someone who enjoyed himself extremely—and who saw things. And as you heard him lecture, you had the feeling that you got a glimpse of what he saw. There was a great beauty about his lectures. Unfortunately, when you came home and tried to work out the notes you had taken, you found that there was a lot to be filled in—you hadn't really seen what he had seen."

After delivering his beautiful lectures, Herglotz, who was a bachelor, would leave the hall without speaking to anyone and go alone to his house. He had almost no contact with students other than Hans Schwerdtfeger and Hanna Mäder, whom Schwerdtfeger later married. But he sometimes dropped in at Busemann's rooms, having learned that the son of the Krupp director, unlike most students, served very good wine.

Hermann Weyl was Hilbert's successor on the faculty. His coming in the fall of 1930 had been preceded, as in 1922, by extended debate with himself and with Hella. In addition to the economic and political situation in Germany, he had been concerned about his own fitness for the position. Perhaps he was too old. He had brought up the name of young Artin, but the suggestion had been rejected in Göttingen. Only Weyl should succeed Hilbert.

Weyl was the mathematical hero of many members of the younger generation, including Friedrichs. Others, like Lewy, were repelled by his literary and personal approach to mathematics and by his philosophical aspirations. They complained that one had to scrape away the words to find out what he was saying. His mathematical papers and books do convey a quality of personality rare in mathematical writing. His great influence came through them.

As a teacher, Weyl was completely different from his famous predecessor and never developed "a school" of his own. There was something tremendously impressive but remote about him. Even his most elementary courses were for only the most gifted. When it came his turn to give the calculus lectures for the beginners, he presented the subject entirely from the point of view of the Intuitionists.

"A very interesting experiment for the lecturer and three of four students," Courant later wrote, "but less useful for the others."

Emmy Noether was not a full professor, but she contributed importantly to the mathematical atmosphere of Göttingen during this period. She and her students, few in number and many of them foreign, represented the trend toward abstraction and generalization which was to become more and more dominant in mathematics during the coming years.

She was a very poor lecturer, writing on the board and wiping out almost immediately what she had written. She spoke quickly and sometimes condensed many syllables into one or two. To Friedrichs it seemed that her speaking never quite caught up with her thinking. "I have no doubt she had a very clear understanding of what she was saying," Hans Lewy told me, "but she didn't have a clear idea of what she was going to say."

She was devoted to her students, who came to her with all their problems, personal as well as mathematical. She was especially popular with the Russian visitors; and when they began to go around Göttingen in their shirtsleeves—a startling departure from proper dress for students —the style was christened "the Noether-guard uniform."

Much of the social life in Göttingen depended on the parties which the professors gave at various times during the year. These were characteristic. Landau's parties were intellectual tests, to which Hilbert never came. Games were played, there were winners and losers. "Hilbert didn't like the premises," Courant explained to me. Herglotz, naturally, gave no parties. Emmy Noether was famous for her "children's parties," to which Hilbert did come. The Weyls hosted a tea dance on a Saturday afternoon— very elegant and formal with many pretty girls present. At the Courants' house there was an unending succession of musical evenings, to which some students were always invited.

I asked Courant if in his new official role as the director of the Mathematics Institute he ever had any difficulty with the other professors.

"No—mostly they were glad to have someone else take over the work, and things were going the way they wanted them to go," he replied. "But I am afraid I was a little bit autocratic. I was always inclined to be as autocratic as the world permits. Only Weyl sometimes objected to what I did. He wanted to be consulted about things; and one time we sat down and had a talk. But otherwise—no."

The new institute was generously staffed in accordance with the government's agreement with the International Education Board. There was an *Oberassistent*, a tenured and salaried official who handled many of the administrative duties. This position was held by the indispensable Neugebauer. There were also ten full-time assistants, all actual or prospective *Privatdozenten*, and a varying number of part-time assistants according to the needs of the *Praktikum* and other activities. In this way, it seemed to Courant, the old "family life" of the mathematics group in pre-war days was being preserved in spite of the much greater number of students since the war.

"[But] it was really restricted to the comparatively small, though by no means exclusive, top group," he later wrote. "This class division could not fail to create among the less fortunate majority the psychological atmosphere of an inferiority complex, which so easily becomes dangerous for scientific situations. . . ."

An American who came to the new institute in Göttingen in 1931 was Saunders MacLane. Up until the time of Birkhoff, MacLane told me, it had been considered impossible for an American to get a proper mathematical education without going abroad. Germany was the place to go, and Göttingen was the most frequent choice. Of 114 American mathematicians who took foreign degrees between 1862 and 1934, a total of 34 received theirs in Göttingen. The second most popular universities were Leipzig and Munich with 9 each. In the early years the attraction was Felix Klein, and five future presidents of the American Mathematical Society were his students in Leipzig or in Göttingen. Later, of course, the attraction was Hilbert. By 1931 not many Americans any longer went abroad for their degrees, but Göttingen was still a popular place for post-doctoral study.

"There was no place like it in the world," MacLane recalled. "It was a real intellectual center. There was heavy excitement going on. There was somehow the feeling that this was the real stuff. That this was the center of things. Everybody talked about mathematics all the time. It was the first real top-class mathematical center that I had ever been at. It was much better than Chicago had been when I had been a graduate student— infinitely better than Yale, where I had been an undergraduate. As I look back now, of course, I have seen many other places that are just as lively and jumping—today in this country Berkeley or Chicago or Harvard— there are many—but it was the first one. It was the real thing."

I asked MacLane how the political situation in Germany had looked to him in the fall of 1931.

"People did not seem very concerned," he said. "Things were always in disorder, but they accepted that. Different people, of course, had different views. My impression that first year was that probably Hitler shouldn't be taken too seriously. Politics in Germany seemed a great big mess. I distinctly remember buying a pamphlet that was labeled 'The 27 Parties of Germany.' There *were* 27 of them, and the NSDAP—the Nazi party—was just one."

FOURTEEN

BY THE END of the academic year 1931–32, the full impact of the worldwide depression was beginning to be felt in Germany. Stringent economy measures were adopted, and university faculties were ordered to dismiss most of the younger assistants.

The members of the mathematics and natural science faculty objected strenuously. Not only would it be unfair to struggling young scholars, who had meager incomes at best; but it would also bring to a virtual standstill the activities of the various institutes. Instead, they proposed that professors pay the salaries of a number of the people affected out of their own pockets.

"The battles this caused with the faculty [as a whole] still make me shudder," Max Born, who was the dean of the science faculty, later recalled. "In the course of an interminable meeting we won with a considerable majority. But those who were outvoted displayed an animosity we had never before experienced. . . ."

It was at this time that Courant received an invitation to lecture in America during the spring and summer of 1932. The work of Americans and their interests had become increasingly prominent in mathematics. Several of them served on the editorial board of the *Zentralblatt für Mathematik*, the new review journal recently founded by Springer. Born and Franck as well as many of Courant's other friends had already been in the United States, and he had heard their experiences. He was "quite curious."

Since Runge's death in 1927, the growing Courant family had moved into the big Runge house on Wilhelm-Weber-Strasse. This was the avenue favored since the turn of the century by the Göttingen professors for their homes. There were now four children, all of whom were cared for and supervised by a much loved housekeeper named Martha Meyer. The household also included Hilde Pick, a cousin of Courant's, who served as his secretary at the new institute. When Courant learned he was going to America, he immediately invited Saunders MacLane to come and live at the house also in exchange for dinner-table lessons in English. Just before he left on his trip, he made still another addition—a young woman student named Charlotte Woellmer, now the wife of Fritz John, a professor at the Courant Institute.

Shortly before Courant died, I talked to Fritz and Charlotte John in their home in New Rochelle; and they told me how he had helped them to continue their educations in the desperate days of the Depression. They had already been friends when Fritz had set out for Göttingen with very little money but the optimistic hope that he would be able to obtain some sort of scholarship there.

"I guess I didn't know how to go about such things," he reflected. "Anyway I was not very successful. But Courant found out about my case and he helped me to get a scholarship. Then he continued to be interested in me. He was always inviting students to musical evenings at his house. I was not musical, but I went anyway. Then fortunately Charlotte arrived, and she became a member of Nina Courant's singing group. But Charlotte had even more financial problems than I."

Charlotte explained to me that she had been able to come to Göttingen only by exchanging room and board with a girl from Göttingen who wanted to live in Berlin.

"Unfortunately—it was very sad—before the end of the semester this girl became ill and died. On the last day of school I met Courant in the hall at the institute and he said, 'Goodbye—see you next semester.' I said, 'I doubt it.' He said, immediately, 'Why? What is the difficulty?'"

When she described her situation to him, Courant suggested that she come to lunch and they would try to see if something could be done. After lunch he said to her, "I am going to the United States soon. How would you like to come and live here at our house for a while?"

"Courant tried—he always went out of his way to help people and to draw people into mathematics," Fritz John said. "Many people—most professors, I think—did not make the effort. There were very few who really had a program of furthering mathematics. They just gave their lectures. Courant had more vision than others in this respect, and he felt a kind of moral obligation to create conditions that would help people."

During the increasingly hard times, a number of informal assistant-ships came into being at the institute in addition to the official ones funded by the government. Often duties were vague or non-existent. Courant once gave a student a stipend because he thought the young man was on the verge of a nervous breakdown and needed a skiing vacation. He also contrived to have some students work without pay. One of these was Busemann.

"Normally I would have become an assistant after I got my degree," Busemann explained to me, "but this was of course the depression time by then; and so, since my father had money and Courant knew it, he asked my father if he wouldn't support me and make it possible for some other youngster, who did not have money, to become a mathematician."

In the early spring of 1932, as Courant prepared for his trip to America, he was concerned about the academic futures of some of his older students. Although Friedrichs was by this time a professor at the technical institute in Braunschweig, Lewy was still a *Privatdozent*; and as times got harder, there was beginning to be evidence of anti-semitism in appointments at German universities. Neugebauer's future was another concern. That young man had chosen a field of interest for which no university in Germany maintained a professorship—the study of ancient mathematics. Since 1923 Courant had managed in various ways to support Neugebauer so that he could continue what Courant—in spite of his own quite different interests—considered important pioneering work. He was currently *Oberassistent* and editor of the *Zentralblatt* and other Springer journals, including a journal for studies in ancient mathematics which Springer had founded for him. He was also a *Privatdozent* for ancient mathematics. With these various sources of income, Neugebauer, now married, managed to live quite comfortably. When he was offered a professorship in Darmstadt, he refused it because, as Courant explained to me, "It would have defeated him." Still, Courant thought, Neugebauer should have a position appropriate to his superior abilities. Perhaps, even during a depression, there would be a place for him as well as for Lewy in the land of millionaires.

During the semester that Courant would be away, he arranged that Oswald Veblen would give lectures in Göttingen. Next to Birkhoff, Veblen was the most influential force in American mathematics. Although not of Birkhoff's stature as a mathematician, he had a personality which was much more appealing to Europeans. ("Veblen combines the best qualities of an American with the best qualities of an Englishman," Hardy used to say.) Courant thought enough of Veblen to let him use his first precious automobile, a Röhr, which he and Nina had just recently learned to drive.

Courant left Germany immediately after a presidential election which pitted the 84-year-old von Hindenburg against four other candidates, one of whom was Adolf Hitler. He arrived in America late in March 1932, when people were more concerned about the recent kidnapping of the Lindbergh baby than about the fact that Hindenburg had failed by 0.4 per cent to obtain a majority of the vote and would have to submit to a runoff election between himself and Hitler.

The American lecture tour took Courant from Columbia in New York City to the Ivy League colleges, then to Chicago and the great mid-western universities, then to the west coast, the California Institute of Technology, Stanford, and the University of California at Berkeley. Everywhere he went he met men and women with earlier connections to Göttingen. The head of the Rockefeller Foundation, with whom he had lunch shortly after his arrivel, was Max Mason, who had taken his Ph.D. with Hilbert. Maria Goeppert-Mayer, a Göttingen girl who had married an American, helped him with the English of his first lecture at Columbia. At Cornell he met Virgil Snyder, one of Klein's Ph.D.'s, who had married a Göttingen girl; and at Princeton he found numerous physicists who had been in Göttingen during the 1920's as well as Solomon Lefschetz, who had given lectures there the year before. At Yale there was Einar Hille, who had also recently spent some months as a visitor in Göttingen. At Harvard he met Birkhoff again and Kellogg, who had been a student of Hilbert, and William Fogg Osgood, a student of Klein. At Brown he saw R. G. D. Richardson, who had been in Göttingen during his own student days, and R. C. Archibald, who came regularly to Göttingen each summer to purchase lecture notes and books. At Chicago the department head was G. A. Bliss, who had studied in Göttingen during the age of Hilbert and Minkowski. On the west coast Courant renewed old times with von Kármán at the California Institute of Technology; and going up to see the telescope on Mt. Wilson, he found Walter Baade, who had been Klein's assistant and the director of the observatory in Göttingen. At Berkeley the head of the mathematics department was M. W. Haskell, who had taken his Ph.D. with Klein and had made the English translation of Klein's Erlangen Program. Courant could not help feeling at home.

"The people are extraordinarily nice," he wrote to Neugebauer. "I am frightened sometimes when I am overwhelmed with attention by people here, for I have hardly paid any attention at all to those same people in Göttingen."

America did not disappoint him by failing to produce millionaires. His letters are full of "the big people" he has met. The words appear invariably in English in the German of his letters. He came over on the boat with an American banker who turned out to be a friend of the elder Busemann. At the luncheon given for him by Flexner he met the head of Macy's ("the largest American department store"). Someone else introduced him to a member of the House of Morgan ("the greatest American banking house"), and he tried—unsuccessfully—to "wangle" money out of him for the Mathematics Institute. In San Francisco, through a banker friend of Carl Still, he met A. P. Giannini, the president of the Bank of America, which until recently had been the Bank of Italy. He was very impressed by this Italian immigrant's son, who had started out as an errand boy in a produce business. He and Giannini had a long talk about the situation in Germany.

"It is simply striking how optimistic about Germany and how relatively adverse to America this man and others of the business people here are in regard to the long-term situation."

Throughout his journey Courant himself nervously followed political events in Germany. In Princeton, with Lefschetz, he listened to the results of the runoff election, in which Hindenburg polled 19,359,642 votes to Hitler's 13,417,460, and discussed with Lefschetz anti-semitism in America. When the Brüning government fell shortly after the election, he begged Nina to save all the newspaper reports for him. Three weeks later he complained, "I still do not understand politics in Germany at all!"

Of the American colleges and universities he visited, he was most impressed by the "old and high" intellectual level at Harvard. But the physical accommodations for mathematics were dreadful. An inadequate library. Bad lecture halls. Impossible blackboards. Everything separated. No applied mathematics. It was obvious that "there was never anyone like Klein to see to the improvement of these conditions."

Princeton impressed him by its luxurious accommodations.

"We try to *Americanize*," he wrote to Neugebauer, "while Princeton tries to surround itself with German *Gemütlichkeit*. In the professors' clubroom . . . there are even newspapers and magazines for entertainment!"

The American university which charmed him most was Berkeley. He wrote at length to Nina about how "nice and friendly" the people were and what a great interest there was in music, "although, strange to say, Debussy and Tchaikovsky play a great role." He found the intellectual standards of the university not very exacting. Although there was a great deal of interest in his lectures, he missed the "resonance" he had felt at Harvard and Princeton.

I asked him once if on this trip he had ever considered remaining in the United States. He said yes, "fleetingly"—at Berkeley.

In the course of his travels Courant did not forget that he was looking for places for Neugebauer and Lewy. In spite of "the really fantastically increasing hardship" in America, he was certain that Neugebauer, whose work seemed to be known and admired, would have no difficulty in obtaining a suitable position. Placing young Lewy would be more difficult, although it was not excluded. The general economic situation in America at the moment seemed to him worse than that in Germany.

"Probably the main difference is the existence of large material reserves everywhere here . . . and, generally speaking, a more stable attitude on the part of the population, who have gone through long periods of economic well-being. But for the rest, there are the same phenomena. . . ."

Three hundred mathematicians and physicists—"some of them quite good"—were going to be unemployed the next year.

While he was in Chicago, Courant heard about a young Ph.D. from the University of Chicago who was planning to spend a year as a National Research Fellow in Germany. He promptly offered to support Edward J. McShane in Göttingen if McShane would translate the Courant calculus books into English. McShane had had only "half a summer's course" in German to prepare himself for a pre-doctoral examination in the language. But times were hard. In the year since he had received his degree only one job opportunity for a mathematician had come to the notice of the head of the mathematics department at Chicago.

"That was from some obscure teachers' college in the sticks, very poor paying even for those times; and Bliss said he was going to do me the favor of not recommending me," McShane recalled. "So I had the brass to accept Courant's offer."

Courant had come to America with a great curiosity about the new Institute for Advanced Study which Abraham Flexner was setting up in Princeton. One of the first appointments there had already been offered to Weyl. Courant knew that two telegrams lay on Weyl's desk in Göttingen, one accepting and one refusing Flexner's offer, and that regularly each morning Weyl debated afresh with Hella the sending of one or the other.

"My first impression was that one should advise Weyl to accept," Courant wrote to Nina. "But [since being here] I have become somewhat more doubtful, in spite of all the difficulties in Germany. . . ."

Late in the summer of 1932—in the middle of an American presidential campaign in which Franklin D. Roosevelt was opposing the incumbent, Herbert Hoover—Courant returned to Göttingen. Weyl refused Flexner's offer; and the winter semester 1932–33 began in Göttingen on November 1, 1932, with the famous faculty intact.

McShane remembers that, in the fall and winter of 1932–33, politics was a favorite conversational topic at the Bahnhof Restaurant, where he and Saunders MacLane regularly ate with the *Privatdozenten* for mathematics. He also remembers that although he and his wife spent the New Year of 1933 in Berlin, they heard nothing about the Nazi riots that were taking place in the capital at that time until they received newspaper clippings about them from their worried families in America.

The Papen government had fallen in December, and in January the Schleicher government also fell. President von Hindenburg went back to Papen and instructed him to explore the possibilities of forming a government under Hitler "within the powers of the constitution." On January 30, 1933, Hitler became chancellor of Germany.

I asked Courant and many others to tell me their reaction, how they felt and what they thought about the future of Germany at that time. Most of them said that they really do not remember any longer—that they have tried to forget and, besides, "so much has happened since then." Even the American MacLane told me he felt he had suppressed the memory of the time. He did recall that after the Reichstag fire on February 27 and the new elections people began to get very nervous. "Immediately there began to come all sorts of regulations, talk about things that would be done at the universities. And then it was unpleasant. At that time I became aware of Courant's particular position, because the Nazis, as far as I can recollect, were especially out to get Courant, as compared to other professors."

Brown shirts and swastikas suddenly began to appear in mathematics lecture halls. The wearers of these were not members of the "in group," but several of them were good mathematicians. Werner Weber, one of the most active of the pro-Nazi students, had been a *Privatdozent* since 1931. The 20-year-old Oswald Teichmüller was extremely gifted. The Nazi sympathies of these and others came as a complete shock to Courant and the young mathematicians of the "in group."

At lunch in the Bahnhof Restaurant, a box was placed in the center of the table and anyone who mentioned politics was fined.

In those nervous days in the early spring of 1933, Courant became possessed with the idea that he must finish the second volume of Courant-Hilbert. Since 1925 he had been trying to work on the book with Friedrichs when that young man was in Göttingen. But somehow he hadn't been able to bring himself to pursue the project. Constantly he had pushed it off. Then in 1930 Friedrichs had gone to Braunschweig as a professor. He continued to visit frequently in Göttingen; and on those occasions Courant would lead him up to a tiny room on the third floor of the new institute building, a retreat which he pretended that no one else knew about. He would then bring out some new section which he had written up—some idea that had caught his interest—and discuss it with Friedrichs.

"He always wanted me to work with him, to read what he had written and talk with him about it, make proposals. He could not seem to work alone. But in all those years, from 1925 on, nothing really happened."

Friedrichs smiled at the memory.

"If it was true, as people said, that Courant's assistants wrote his books, I would have written Courant-Hilbert II."

At the beginning of March 1933, Courant arranged that Friedrichs and Franz Rellich would accompany him and his family to Arosa during the spring vacation. They would ski and work, and they would *finish* the second volume of Courant-Hilbert!

Courant was aware of rumors which were being circulated in Göttingen that he was planning to "flee" to Switzerland. He considered cancelling the ski-work trip because, as he explained to Franck, "it seems unfeeling of me to sit calm and free abroad at this time." But his children needed the fresh air and sunshine of the mountains. Friedrichs and Rellich were already waiting in Arosa. He decided to go ahead with his plans.

The sequence of events, the emotions, and the reasoning from which Courant and others acted during the next few months were to fade in the coming years, what was actually known at the time confused with what became known only later, what should have happened with what did happen. For this reason I have tried to limit my account to those documents which Courant took with him to the United States in 1934 and which he turned over to me in 1971, a few months before his death. He could not bring himself to go over them in detail but merely glanced at them, saying he found it impossible to believe that he had written what he saw there.

While Courant was in Arosa, Hitler pushed through the Enabling Act, which gave the government power to issue decrees independently of the Reichstag and of the President. A nationwide boycott of Jewish business and professional men was announced for April 1.

On March 30 Courant wrote to Franck to ask his advice about whether he should return immediately to Göttingen. He was sure that if he could stay in Arosa as planned, he would be able to finish at least half of the second volume of Courant-Hilbert. And yet—he had been somewhat alarmed the day before when he had heard through the servants that people were saying in Göttingen that he would not be back.

"Incidentally, it is my impression from newspaper reports that things have been going peacefully and quietly in Germany," he wrote to Franck. "From the first I have been appalled to read reports of how such people as Einstein have made statements and what lies and 'latrine rumors' have been utilized to make Germany's internal situation a butt for general political agitation abroad. If I had known Einstein's address, I would have written to him. . . ."

Einstein, who had been in America for the past few months, had been making a number of widely publicized statements deploring "brutal acts of violence and oppression against persons of liberal opinion and Jews . . . in Germany [which] have aroused the conscience of all countries remaining faithful to ideals of humanity and political liberties." He hoped that world reaction would be sufficient "to preserve Europe from regression to the barbarism of past epochs." On March 29, the day before Courant wrote to Franck, the government in Berlin had announced that Einstein had inquired about taking steps to renounce his Prussian citizenship.

"Even though Einstein does not consider himself a German," Courant wrote, "he has received so many benefits from Germany that it is no more than his duty to help dispel the disturbance he has caused. Unfortunately, as I see from the papers, a reaction to these events has set in. . . . I very much hope that it will be possible to deter the intended boycott [of the Jews] at the last moment. Otherwise I see the future very black.

"What hurts me particularly is that the renewed wave of anti-semitism is . . . directed indiscriminately against every person of Jewish ancestry, no matter how truly German he may feel within himself, no matter how he and his family have bled during the war and how much he himself has contributed to the general community. I can't believe that such injustice can prevail much longer—in particular, since it depends so much on the leaders, especially Hitler, whose last speech made a quite positive impression on me."

Franck's reply is undated, but it seems to have been written on March 31. He advised Courant to try to quiet the rumors about himself by returning to Göttingen immediately.

"It is also possible that any day decisions may be necessary which can only be really understood on the spot."

April began with the "non-violent" boycott of Jews, although officially it had been withdrawn by the government. Hilbert, who was still an inveterate gardener, picked a bouquet of flowers and sent them to his Jewish doctor.

"It will not be long before the German people find Hitler out," he said, "and then they will put his head in the toilet."

On April 7 Courant, who had returned to Göttingen without his family, wrote reassuringly to Nina. On that same day the government announced a series of laws—the *Reichsgesetze*—for the "restoration" of the professional civil service. The intent of the new laws was to remove from government employ all persons of "non-Aryan" descent with certain exceptions, mainly military, and also to remove persons who were considered politically unreliable. Since university professors were civil servants, the new *Reichsgesetze* applied to them as well as to judges, to officials and employees of innumerable public and semipublic agencies, and to many others.

In Göttingen, Courant and Franck could take comfort in the fact that non-Aryans who had fought for Germany or her allies were specifically exempted; unfortunately Max Born, who had done war-related research, was not so protected. How far would the government go in carrying out the *Reichsgesetze*? In the second week of April 1933, it seemed that the violent anti-semitism simply could not last.

But—demanded James Franck of his friends—didn't the very fact of the new laws, removing as they did German citizenship from many of Germany's Jews—declaring that their children were *not Germans*—require response and protest on the part of Jews?

FIFTEEN

THE University of Göttingen had a certain tradition of political protest which was as old as its mathematical-physical tradition. During the time of Gauss, seven famous professors, including the physicist Wilhelm Weber but not Gauss, had protested the repeal of the liberal Hanoverian constitution. Their subsequent dismissal had aroused a storm of indignation throughout Europe.

In 1933, almost a hundred years later, Courant, Born, and Franck considered a joint protest in the spirit of the Göttingen Seven. Courant and Born thought that such an act might precipitate action from the government which would not otherwise be forthcoming, but Franck insisted that he wanted to resign. They debated whether they were not morally obligated to follow him. Although Born was more vulnerable under the law than Franck, he drew back. He and Courant agreed that they should remain at their posts and do all in their power to save what they had helped to create in Göttingen. Courant, even more than Born, felt a personal responsibility to remain because the Mathematics Institute had been so recently built as a result of his personal negotiations between the Americans and the German government. He was also inclined to postpone irrevocable action.

For a week the discussions among the three professors and Neugebauer continued. Then, on April 13, several individuals at other universities in Germany were summarily placed on leave. On Easter Sunday, April 16, Franck came to his decision. He wrote to the minister asking to be relieved of his duties. In a letter to the rector of the university, he explained his action:

"We Germans of Jewish descent are being treated as aliens and enemies of our homeland. It is required that our children grow up with the knowledge that they will never be allowed to prove themselves as Germans. Those who fought in the war are supposed to have permission to continue to serve the State. I refuse to avail myself of this privilege, even though I understand the position of those who consider it their duty to remain at their posts."

He intended, he said, to try to continue his scientific work in Germany.

Angry response to the announcement of the resignation of Franck, a Nobel laureate, came first from forty-two members of the faculty, who

142

issued a statement condemning him for giving the foreign press material for anti-German propaganda. Many of those who signed were the same people from agriculture and forestry who had bitterly opposed the scientists' proposal that the professors pay the younger assistants out of their own salaries. Rumors began to circulate that Born and Courant were part of a plot to "sabotage" the new government and that they had had Franck alone resign simply for "tactical" reasons.

Two days after the faculty statement, without any direct notification to those involved, an announcement appeared in the local newspaper on April 26 that, in accordance with the new *Reichsgesetze*, six Göttingen professors were being placed on leave until further notice. Four of the six were from the scientific faculty: Felix Bernstein (who was in the United States at the time), Max Born, Richard Courant, and Emmy Noether. The headline on the little story announced that more would follow.

Courant heard the news when a friend who had seen the paper telephoned him.

He immediately sent a letter to Harald Bohr. The optimism of the first weeks of April was gone. He wrote Bohr that he was convinced—as was everyone else in Göttingen, including the local authorities—that the "leaves" would shortly and irreversibly be converted into dismissals.

"I have been harder hit by the turn of events and less prepared than I should have thought," he confessed. "I feel so close to my work here, to the surrounding countryside, to so many people and to Germany as a whole that this 'elimination' hits me with an almost unbearable force."

His personal plan was to remain quietly in Göttingen—financially this would be possible for a number of months—and to do his best to complete the second volume of Courant-Hilbert. "Only in this way can I maintain the energy and the moral justification for continuing in my scientific career." As for the more distant future, positions would have to be found for himself and the other dismissed professors. The best hope seemed to him to lie in America with Flexner and his many connections, but he did not feel that he should personally approach Flexner.

"How hard it is for us to make use of your friendship in this manner, we need not assure you," he told Bohr. "I hope it will never be necessary to show you that at any time we would be prepared to do the same for you."

It may have been this letter from Courant to Bohr which Neugebauer remembers carrying to Hamburg so that he could place it directly in the hands of the Danish consul there.

"The very first moment was physically extremely dangerous," he told me. "One didn't know what the S.A. would do. From the time they took over, there was brutality."

Support for Courant came immediately from many friends and former students—in some cases even from people from whom he did not, perhaps, expect support. One of these was Hellmuth Kneser, who had been his first assistant in Göttingen. Kneser was a professor in Greifswald and was known to be impressed by the new government, which "despite everything" had—in his opinion—brought about a revolution with quite unrevolutionary peace and order and had been able to unify the *Reich*—something which neither Bismarck nor the Weimar government had been able to accomplish. He was considering joining the National Socialist Party —"not," as he said, "for the sake of influence or advantage—if I had desired that, I would have joined earlier—but so as not to be farther removed externally than internally." He wrote Courant, "very startled," the day the news of the "leaves" appeared. "I don't understand at all what has been done, unless it is intended to prevent disturbances [at the university] in the future."

Courant replied with an extensive letter. He had not yet received any official notification of his status. He was preoccupied with "the reason" for his dismissal, since he should have been exempted from the *Reichsgesetze* because of his wartime service. He mulled over the details of his short-lived political career but could not imagine that the facts of this could be sufficient to provoke a leave order. He recognized that he and the scientific faculty were "instinctively abhorrent" to various groups. There was probably a certain amount of unconscious jealousy involved. Wild stories and rumors had developed which could be summarized by the description of the Mathematics Institute as "a fortress of Marxism." Much of the hostility had centered upon his person, because it was he who had been responsible for the erection of the institute. And then there was the case of Franck—"I can imagine that it was Franck's resignation which actually provoked the ministry's action."

The situation at the Mathematics Institute became increasingly volatile. Kneser, Courant knew, had personal ties with an influential party member in the ministry, who might be able to help. The day after his first letter, he dispatched a second letter to his former assistant.

"Not only are the students apparently determined to try to prevent Landau and Bernays from lecturing, but they are also attacking Neugebauer as 'politically unreliable'—that is, communistically oriented. The dean has stood up for Neugebauer, but it does not look as if Neugebauer can endure the pressure. He now holds—as my representative—the position of director of the institute; but I am afraid that he will give it up if the students don't withdraw their threat to boycott him. Since Weyl is not a strong and stable personality and since Herglotz cannot be considered for the directorship either, I see the future of our institute as very dark. . . .

"It is a pity to think what treasures are going to be destroyed in this way after more than ten years of work at reconstruction. It pains me most to see what senseless damage will be done to Germany. Only look at the Americans and other foreigners who are now getting ready to break off their studies and go home! In any case, the spirit of our institute has already been destroyed. Ugly signs of opportunism have become evident. I am much afraid that quite apart from what has happened to me irreversible actions have been taken. . . . Still, if the matter of Neugebauer could be straightened out, I would be very happy."

After his letters to Kneser on April 28 and 29, Courant fretted constantly about counteracting the rumors that were being circulated about him. According to one of these, he had carried a red flag during the days of the 1918 revolution and had disarmed troops returning from the front. He was convinced that such unfounded stories, or at least the hostile attitude from which they stemmed, were responsible for his being placed on leave. Although he had not yet received official notice of his status, he made up his mind to approach officialdom.

Seeking a colleague unaffected by the *Reichsgesetze* who could present his version of his activities to the administration of the university, he settled upon Prandtl. The professor of mechanics was generally considered a somewhat naive man. But during the wild week following the announcement in the paper, he had acted with courage and decision, firing one of his assistants when he had discovered that the man was an informer for the Nazi forces at the university.

It was decided that Courant should write a letter to Prandtl setting out the facts of his political activities after the war. Prandtl would then present the letter to the *Kurator*.

In this letter, dated May 1, Courant covered essentially the same points he had made in his first letter to Kneser, with the exception of the fact that he also took up a charge that he had "jewified" the Mathematics Institute. He pointed out that Lewy (who had already left Göttingen—the first to go) was the only Jew whom he personally had appointed to an assistantship.

"I don't have to say anything about my attitudes," he concluded. "Better than words, the facts testify. I have served the general interest in all my activities, from which the university has surely benefited."

Since the previous Thursday, when the announcement of his leave had appeared in the paper, Courant had not gone to his office in the Mathematics Institute. From a practical point of view there had been almost no disruption in activities, since he and Neugebauer had always worked closely together. But he continued to worry about how long Neugebauer would be able to handle his increasingly difficult role.

Over the weekend Neugebauer was asked to sign the required oath of loyalty to the new government. He refused and was promptly suspended as *untragbar*—a word which means in German, quite literally, "unbearable." Although he continued to receive his salary as *Oberassistent*, he was forbidden to enter the institute building.

As May and the new semester began, for the second time within a week, the Mathematics Institute of Göttingen was without a director.

Although initially Courant had felt that he should not personally approach Flexner for help in obtaining another position, he now wrote the American a long letter.

There was, he explained, a small and constantly diminishing number of people who still remained aloof from the new nationalistic ideas and enthusiasms. "The government possesses a power and ability to act such as probably has never existed in the world before." The picture of the German universities which Flexner had presented in his famous book was no longer a true one. "To my distress I foresee that the flowering of our scientific group here is past."

Neither he nor Neugebauer had been able to make any personal decisions about the future. He was continuing to try to finish the second volume of Courant-Hilbert, and Neugebauer was continuing his historical researches.

"Quite frankly, I am writing you to ask whether you can help us. You may be sure that I do not have in mind one of those brilliantly distinguished positions at your institute, but would be perfectly satisfied with some modest place. On the other hand, I believe quite objectively that I could again be productive and of use in new surroundings. The same naturally applies to Neugebauer."

On May 5, nine days after the announcement of their leaves in the newspaper, the six affected professors received official notice of their status.

Courant renewed his efforts to set his personal record straight. In response to the official notification, he composed a lengthy description of his activities and sent this for transmission to the ministry with a personal letter to the *Kurator*: "I repeat that my letter should not give the impression that I am asking for mercy. I am interested only in not being stigmatized as a bad German, or as a negligent civil servant."

Max Born, shortly after receiving the official notice of his leave, left Göttingen for the Dolomites, where he and his family had already planned to spend their summer vacation. On May 24 he sent his resignation to the ministry, expressing his basic agreement with Franck's position as expressed in Franck's letter of resignation.

Franck himself remained in Göttingen, conducting a physics seminar in his home and trying to help his assistants and students to place themselves outside Germany.

Courant's own activities and statements during the period often appear contradictory. From the moment the leaves were announced in the newspaper, he was convinced—as he wrote Harald Bohr the same day—that they would inevitably become dismissals. At the same time, during April, May and June, he put up a determined fight to reverse his leave.

Perhaps Max Born, writing to Einstein from the peace of the Dolomites, had the clearest understanding of both Franck's and Courant's actions at this time:

"Franck is resolutely determined not to go abroad while he has the slightest prospect of finding work in Germany (though not as a civil servant). Although there is, of course, no chance of this, he remains in Göttingen and waits. I would not have the nerve to do it, nor can I see the point of it. But both he and Courant are, in spite of their Jewishness, which is far more pronounced than in my case, Germans at heart."

After Neugebauer was declared *untragbar*, Hermann Weyl became acting director of the Mathematics Institute and immediately joined in the efforts to have the leaves of Courant and Emmy Noether lifted.

In Courant's case, Weyl thought that the best approach would be to try to obtain official documentation of his services to Germany during the war. By the middle of May letters from old comrades and co-workers began to arrive. Some were obviously brush-offs, but many were sincere testimonials.

Most heartwarming was the letter from the major under whom Courant had first served in 1914 and whose life had been saved in the Argonne by the bravery of Ernst Courant. Now a lawyer in a small country town, this man wrote a sympathetic personal note and enclosed a detailed account of the wartime activities of both brothers while they were under his command. He described Ernst as "fearless and extraordinarily brave" and Richard as "an absolutely reliable soldier and a true comrade . . . helpful and tireless. . . . His quiet sense of duty made him an example to officers, non-commissioned officers, and men."

These letters, as well as typewritten copies of newspaper reports of Courant's post-war political activities and a reminder of his work in getting out the vote in the plebiscite of 1921, were sent by Weyl to the ministry on May 18.

At about this same time Hellmuth Kneser had an opportunity to approach the ministry's representative, Theodor Vahlen, about Courant's case.

Vahlen, a mathematician who had written the first German textbook on ballistics, had been a student of Klein in Göttingen. Once, on the occasion of the Day of Nationality, when he was a dean at Greifswald, he had personally ripped down the black, red, and gold flag of the Republic and hoisted in its place the old black, white, and red of the Empire. When Kneser had left Göttingen, he had gone to a professorship in Greifswald; and Courant was always inclined to attribute his later political views to Vahlen.

"Kneser was young, and he got some ideas under the influence of this dean," Courant explained to me. "I don't think he ever did anything really bad. He is a good man. A very good mathematician."

In spite of Vahlen's new position in the government, he still came regularly to Greifswald on Saturdays for a lecture. He was occupied most of the day with the local S.A. leadership, but he spent an hour with Kneser talking over the cases of Courant and other professors who had been placed on leave. Kneser reported to Courant that Vahlen had assured him that only established facts would be considered and that every affected person would be informed of the various points of the law which applied to him and would be given an opportunity to rebut them. Vahlen knew Courant only from rumors, "such as that you were a Zionist, which of course I denied immediately." He convinced Kneser of "a serious intent to be scrupulous." The most important thing in the conversation seemed to be the fact that when Kneser said of Courant, "But I know him—I was his assistant for four years," Vahlen replied, "Then you also could be asked to appear."

"So if you think of any point on which I could testify for you," Kneser instructed Courant, "call me or other suitable people."

While Kneser was approaching Vahlen on Courant's behalf, Friedrichs came to Göttingen to see Neugebauer. The two men agreed that they must do something to help Courant. They decided on a statement about his activities as professor and director of the Mathematics Institute. This would be signed by former students and other colleagues with connections to Göttingen and then sent to Bernhard Rust, the new minister for education.

By the time they finished writing, they had a long statement running to almost four double-spaced typewritten pages. Invoking revered and Aryan names, they began by pointing out how Klein and Hilbert had given the great scientific tradition which had developed in the work of Gauss, Dirichlet, and Riemann "a decidedly new turn." It had been they who had called the young Münster professor to Göttingen to carry on their work after the war. Courant, "a student of Hilbert," had devoted himself "with never tiring energy" to this task. His scientific works, particularly the great book written with Hilbert on the methods of mathematical physics, were now classical tools of analysis. In line with Klein's educational ideas, he had reorganized instruction in Göttingen to meet the needs of the unexpectedly great number of students coming to the university after the war. "As we are indebted to Klein for the mathematical *Lesezimmer*, which is available to all students, so Courant has established the great *Praktikum* for mathematical beginners and has extended it through years of work." It was thanks to Courant's initiative and energy that Klein's plan for the building of a mathematical institute had finally been realized— "offering opportunities for Germans as well as foreigners that are not offered anywhere else in the world." His "constant care" had been devoted to more than the welfare of mathematics—they cited the improvements to the physics institute, the building of the mineralogical institute, the extension to the institute for organic chemistry. In the latter part of their statement they turned to the activities of Courant with which they were most familiar—the discovery of talent, the financial aid and support of students, Courant's role as counselor and friend to the young.

"Rarely has a man tried so little to get credit for his achievements in the eyes of the outside world. He has never accepted opportunities offered him to go abroad because he is so closely connected with his work in Göttingen and because Göttingen has become so entirely his scientific and personal home. It is because of his personality that many of us also remember our time in Göttingen as a life in a scientific homeland."

In one of our conversations Friedrichs brought out the faded folder which contains a rather tattered copy of his and Neugebauer's petition, the final list of signatures, and the letters written by many who refused to sign.

On various pieces of paper in the folder, eighty-seven signers are suggested. Of these, twenty-two have been eliminated for various reasons, frequently it seems because they were teaching outside Germany and anything which could be construed as "meddling from abroad" was known to antagonize the government. Of the sixty-five remaining, sixteen did not respond at all. These included one man who had been Courant's assistant for several years. Of the other forty-nine, twenty-one refused to sign but wrote letters explaining their reasons.

There were various reasons—they were no longer close to Göttingen, or they did not know Courant well enough to judge whether the statements made about him were true. One writer explained that he and a colleague had signed immediately and had set about to obtain more signatures for the petition. They had, however, encountered so much negative response and had been so "urgently requested" by their colleagues not to participate that they had become hesitant and "after two days of consideration" had come to the conclusion that they should think the matter over further. There was an admitted feeling on their part that, much as they personally would like to support Courant, they did not want to bring themselves as individuals to the attention of the government. Was the Göttingen *Kurator* personally supporting the petition, they asked, and had any mathematicians who were "positively oriented" to the new government signed? If so—and if there were at least twelve signatures and if Friedrichs and Neugebauer thought it was necessary—they would add their signatures by wire. He was ashamed, the writer concluded, but he had to admit that he was afraid of the consequences if he signed.

Some did not consider a petition a proper vehicle for approaching the ministry. One found parts "disrespectful and offensive"; another said that he was sure "our minister of education, Mr. Rust, will do the right thing." Some thought the petition should be postponed or changed.

"I cannot sign," wrote one man who had been a student in Göttingen, "because I disagree very much with the general tenor of the petition. . . . Courant's stature as a mathematician is uncontested and well known. . . . His political status should be clarified. Proof should be found that his activities in this connection were undertaken for the sake and safety of the university."

"I remember a remark of Klein's to the effect that he did not care for Courant's political outlook and was inclined to think that he had Marxist leanings," another professor contributed. "I want to warn people against signing this petition, which will not help Courant and which may be very harmful to those who sign it."

Others, refusing to sign, said that they would nevertheless be willing to support Courant personally vis-à-vis the ministry. Among these, Carl Ludwig Siegel, by then a professor at Frankfurt, rewrote the petition in a way which seemed to him more precise and to the point and sent his version directly as a personal letter in support of Courant.

Kneser, who also did not sign, thought that a more effective way to help Courant would be to have a small number of colleagues and former students or assistants ("preferably party members before January 30, but they will hardly be found") write directly to the minister and ask to be allowed to testify as to their personal knowledge of Courant's character and his activities. The suggested letter, which was signed by Friedrichs, Kneser and Prandtl, began:

"Each of the undersigned knows Professor Courant as the result of a number of years of close collaboration. To our knowledge in all of his activities he has felt himself a German citizen and a representative of German science and has conducted himself as such. . . . The mathematical facilities of Göttingen, which since 1921 have been developed essentially through his efforts, are of great significance for the scientific culture of Germany and will not without essential damage be separated from his person."

It concluded with a request that the three be heard in person and, if this was not possible, that they be permitted to contribute their testimony regarding Courant in writing.

At the end of two weeks, twenty-eight men, more than a third of them physicists, had agreed to place their names on the Friedrichs-Neugebauer petition. These included such well-known scientists as Artin, Blaschke,

Carathéodory, Hasse, Heisenberg, Herglotz, Hilbert, von Laue, Mie, Planck, Prandtl, Schrödinger, Sommerfeld, van der Waerden, and Weyl.

Neugebauer found it "quite a nice list" as he conveyed it to Friedrichs on June 8—"although there are many not to be seen who should be seen."

"Several of the names on the list are those of people who later were considered to be Nazis or near-Nazis," Friedrichs pointed out to me, "and even at the time some of them were known to be in sympathy with the regime."

Today the judgment of motives and actions seems to him more difficult to make than it did then.

"You see, it is not clear. It is not clear at all."

One of those who refused to sign the petition in support of Courant was Hecke, who had known him since their earliest university days in Breslau and had come to Göttingen and been chosen as Hilbert's assistant on his recommendation. Although Hecke was in sympathy with the purpose of the petition, he felt that it would have either no effect or the opposite effect from the one desired.

It simply did not direct itself to the political mentality of the government.

"And of course Hecke was right," Friedrichs said as he put the old letters back into the folder.

SIXTEEN

THERE WAS no response from the ministry to either petition. Courant sat at home and fidgeted. He tried to concentrate on the second volume of Courant-Hilbert, but the work refused to move forward. Psychologically he was in a worse state than he had been since the war. He leaned heavily on the Bohr brothers and the contact they represented with the outside world. Once they came to Germany, and he and Neugebauer met them on a North Sea beach and enjoyed the pleasure of speaking freely for a whole afternoon.

Weyl still served as director of the Mathematics Institute. Since Hella Weyl was Jewish, his tenure in that office was precarious; but he worked tirelessly, writing letters, conferring with government officials, trying to assure continuity.

Busemann, who was a quarter Jewish and thus still unaffected by the racial laws, remembers being begged by Weyl to remain at the university. "But I had already lived under a dictatorship when I was in Italy, and I had made up my mind to leave Germany and sit out Hitler in Copenhagen."

Fighting to save something in Göttingen, Weyl was still full of self-recrimination that he had refused the offer from Flexner and had doomed his family to Nazi Germany. Courant worried the question, if Weyl left, who would carry on the tradition of Klein and Hilbert? He placed his hopes on Rellich, who, although not a professor, had been Fritz John's examiner; but he feared that Rellich, while personally inoffensive, was too closely connected with the *Juden-Fakultät* of Göttingen to be acceptable to the ministry.

The future of Fritz John, who was half Jewish, worried Courant too. There had been a last minute crisis about John's getting even enough money to pay the fees required for the doctor's examination. To make matters worse, ten days after he got his degree, he had married the equally poor Charlotte Woellmer, who was not Jewish. Marriages between non-Aryans and Aryans, while not yet forbidden, were looked upon with such disapproval that Fritz and Charlotte had told no one about their plans. During the summer they survived on small sums paid by Courant from vague funds he still had. They also made some money from the sale of Fritz's lecture notes, typed by Charlotte. These were purchased by an American professor at a university called Brown which, they assumed from its address, was located on an island off the coast of the United States.

Courant was convinced that Fritz John would be an outstanding mathematician, and in the midst of his own immediate problems he fretted about finding a position for John outside Germany.

At the end of the summer semester, as MacLane and McShane and their wives prepared to leave Göttingen, Courant gave McShane an envelope of money addressed to Harald Bohr and asked him to mail it outside Germany. McShane told me that the money was used by Bohr to support young Jewish mathematicians from Germany who were already by then in Copenhagen.

It was at this same time that Courant received a letter from a former Turkish student, who had heard that he and Franck might accept appointments at a newly reorganized state university in Istanbul. Ten days later, at a meeting in Zurich, the two professors were surprised to be formally approached by a representative of the Turkish government. Courant took a very reserved position.

"I am—though on leave—a Prussian employee and would consider it impossible (either personally or professionally) to enter into negotiations regarding service in a foreign government without informing my superiors," he explained in a letter written later from Göttingen to the Turkish representative. "The most important reason to me, however, is that—if conditions force me to go abroad—I go, not as an embittered emigrant, but as a proud representative of German culture who neither internally nor externally has severed his ties with his homeland."

He was nevertheless "very interested" in the idea of creating "a scientific center of European quality" in Istanbul and eager to help "to the fullest extent of my ability." Later that summer, apparently assured that his current status would not be affected, he made a trip to Istanbul with Franck. On the way—or perhaps it was on his earlier visit in Zurich—he let Flexner know that if Weyl were again offered a position at the Institute for Advanced Study he would accept.

In Istanbul, Courant and Franck found a beautiful but alien city. Hitler's dismissal of so many outstanding scholars in every field had coincided with the desire of Ataturk (the name was not actually adopted until the following year) to staff his new state university with the most distinguished faculty in the world, At first Courant had been attracted by the idea of going to Turkey, and Nina had been very much in favor of it; but observing the political conditions on the spot, he decided that nationalistic Turkey was not for him.

"But why did you even consider going to Turkey instead of going to the United States?"

"It was closer to Germany."

Before the end of summer 1933, Courant received an invitation to spend the coming academic year in England as a visiting lecturer at Cambridge. His official status was still that of a German civil servant—although one on "forced leave"—and he had to obtain official permission to accept the English invitation. This was a delicate maneuver, since the question of the lifting of his leave was pending. Finally, ten days before the beginning of the term, he was informed that—in view of the fact he would not be in Göttingen during the coming academic year—his leave was being expressly lifted; in short, he was no longer on "forced leave" but was simply "on leave" at another university.

Courant left his family in Göttingen. The children continued in school with only a few incidents—in Courant's view, these were for the most part comical rather than tragic. He was more concerned about the indoctrination taking place in the schoolroom and on the playground and the natural desire of young people to be like their peers. A few days after he arrived in England and while he was still staying with Born, who also had a temporary appointment in Cambridge, he took the opportunity to write freely to Flexner about the importance of getting young scientists out of Germany immediately:

"The Nazis have remained consistent only in regard to the so-called Jewish question. . . . They indoctrinate a ridiculous racial theory (the basis of which is anti-semitism) through propaganda of all kinds . . . and it may easily happen that once this poisonous seed has germinated, an atmosphere much worse than that now existing will have been created."

A week later Courant wrote again to Flexner to tell him that although the government appeared to be having second thoughts about the removal of well-known professors—as in his own case and that of Schrödinger—the situation of Jewish professors in Germany was hopeless.

"The following event is characteristic of the course of things. Professor Landau, [to whom the *Reichsgesetze* do not apply in any way], went to start his lectures last week. In front of his lecture hall were some seventy students, partly in S.S. uniforms, but inside not a soul. Every student who wanted to enter was prevented from doing so by the commander of the boycott. [This was Weber, who had once been Landau's assistant.] Landau went to his office and received a call from a representative of the Nazi students, who told him that Aryan students want Aryan mathematics and not Jewish mathematics and requested him to refrain from giving lectures. . . . The speaker for the students is a very young, scientifically gifted man, but completely muddled and notoriously crazy. [From this description Friedrichs recognized Teichmüller.] It seems certain that in the background there are much more authoritative people who rather openly favor the destruction of Göttingen mathematics and science."

Courant's appointment at Cambridge was very specifically for one year only, and almost immediately he realized that in spite of the kindness of the English mathematicians he would not be able to live permanently in England.

"Mit den Engländern I get along very well," he wrote in a mixture of German and English to Franck, who was now in the United States. "All the objections you had regarding my temperamental faults have proved to be unjustified, because I have really taken advantage of the change in environments in order to change my outward attitude to people I do not know as well as I know you. I believe, therefore, that I 'pass' here. Still, if serious prospects develop in America, they will have to be given preference for the long term."

After receiving the Cambridge invitation, Courant had received an invitation to spend a semester at Berkeley. He had had to refuse it, since it was for a shorter period of time; but it had aroused in him the hope that there might eventually be a place for him at the University of California.

Flexner wrote that he was "actively exploring possibilities" for him. Veblen, who was now at the new Institute for Advanced Study, also wrote, "Your friends in America are trying to find a worthy position for you."

In Cambridge, Courant found that enthusiasm for finishing his book very soon began to evaporate. He was overwhelmed by guilt that in spite of freedom from the administrative responsibilities that had excused his failure for so long, he was still not able to write. People who had known him earlier found him "lost" in Cambridge.

At Christmas he returned to an "empty" Göttingen. The only one of the mathematics professors left at the university was Herglotz; the only physics professor, Robert Pohl. Weyl and his family had gone to America.

(Weyl's son Joachim recalls that during the preparations for the move his father went about the house intoning, as he often did, appropriate lines of poetry. "I cannot remember exactly," Joachim Weyl told me, "but there was something about 'if the beast of tyranny takes over his country, he puts the torch to his own house and leaves.' The last line was *zu dienen im Dunkel dem fremden Mann*—'to serve in the dark the unknown man.'")

The sad state of the Hilberts—Käthe Hilbert was almost blind—added to the bleakness of the holidays for Courant. Also Martha Meyer, the housekeeper whom he had hired for Nina when their third child was born— a woman who had been like a mother to all their children—announced that she felt it would be better if she did not remain any longer in the household of a Jewish family. It was rumored that Heisenberg might take Born's place, but that Franck's professorship would be done away with. The most shocking development was that Werner Weber—the assistant who had led the demonstration against Landau—was the new director of the Mathematics Institute.

When Courant returned to Cambridge after Christmas, he was accompanied by Nina and their eldest son, Ernst, as well as by Fritz John, for whom he had managed a small fellowship. But he was still miserable.

"Day and night the thought tortures me what is supposed to become of the family in the summer," he wrote to Franck. "Please do not be angry with me . . . that I am deeply depressed. I know very well that, as in the case of Ehrenfest, all such difficulties actually come from within. . . . I know that I must cope with them."

Weyl sent happy letters from Princeton. In Fine Hall, where Flexner's group was temporarily lodged, German was spoken as much as English. He frequently saw Emmy Noether, who was nearby at Bryn Mawr. He wrote that he was "thoroughly satisfied with [the Institute for Advanced Study], the spirit and the work which is being done."

At the end of January, Courant received a long heralded offer from New York University. He would be given a two-year contract with the possibility that the position might become permanent. The salary was to be $4,000 a year, half of it to be paid by the Committee in Aid of Displaced German Scholars and the other half by the Rockefeller Foundation.

Courant was disappointed. Even in the early 1930's, $4,000 did not look like very much to a German professor who had been receiving a salary equivalent to $12,000. It was, in fact, a $1,000 less than what Felix Klein had been offered by Johns Hopkins University in 1883 to succeed Sylvester. Klein had refused *that* as inadequate and had gone instead a few years later to Göttingen.

Courant's disappointment was intensified by the fact that at the same time he received the offer from New York he also received another semester's invitation from Berkeley. It would pay him $3,000 from the middle of August to the end of December. He was torn again. Veblen wrote from Italy, Franck from Baltimore, Weyl from Princeton, all in favor of New York. "I think you simply *must* accept this offer," Weyl told him sternly. But Courant hesitated. Would it be possible, he wondered, for him to go first to Berkeley and then to New York? His friends assured him that it definitely *would not*.

Courant had no idea where—in New York—New York University might be. On his visit to the city in 1932 he had been aware only of Columbia. He wrote helplessly to Veblen, reverting to German, although by now he was customarily writing in English to the Americans: "Wer sind dort die Mathematiker?" *Who are the mathematicians there?*

Veblen replied that the only mathematician he knew at New York University was an assistant professor named Donald Flanders.

". . . he will be of special interest to you because he played a role in your history. Flanders is a Ph.D. from the University of Pennsylvania who studied in Princeton for a year or two as a National Research Fellow. His field is topology, and he is very zealous about research but has not been correspondingly successful. Early in the academic year he came to see me to get my advice about improving the mathematical situation in New York University. He was very anxious to get them to call in some real mathematicians. I was quite impressed with his unselfishness in the matter, because it was clear from the first that it would mean calling people in who outrank him in every respect, and he was under no illusions on the subject."

Courant wrote immediately to Professor Flanders: "Every kind of information about the general level of lectures, the type and preparation of the students, the facilities of mathematical studies would be very precious to me."

In spite of Courant's own troubles, all through the year at Cambridge he continued to be the person other professors who had been placed on leave turned to for help. A record of their pleas and questions and his responses is contained in a bulky folder of correspondence labeled "1933–34," which he later brought to the United States. Many of his replies were typed by Charlotte John, who came to England in March to join her husband. Courant paid her from funds he had from Göttingen, since Carl Still had insisted that he not return money that Still had already given him for the year.

From Germany that spring came rumors of a disturbing speech by Bieberbach in which the Berlin mathematics professor had applied a current theory of personality types to the practitioners of his own subject. One type was represented by "true Germans"—Hilbert and Klein were examples—and the other, by Frenchmen and Jews.

To support his thesis that "a German essence" exists in mathematical creation, Bieberbach had quoted a statement made by Klein in America in 1893: "It would seem that a strong naive space-intuition were an attribute preeminently of the Teutonic race, while the critical pure logical sense is more fully developed in the Latin and Hebrew races."

One of the great achievements of true German mathematics, according to Bieberbach, was Hilbert's work on axiomatics; and it was most regrettable that "abstract Jewish thinkers" had succeeded in turning this work into "an intellectual variety show."

"Our nature becomes conscious of itself in the malaise brought about by alien ways," Bieberbach had explained. There was an example in the "manly rejection" of Edmund Landau by the students in Göttingen. This man's un-German style in research and teaching was "intolerable to German sensibilities." The important task for "National Socialist science" was to recognize the existence of the "German essence" in works of science and then "to proceed to action"—the nurture of that essence.

At the time of Bieberbach's speech, Courant was already beginning to make preparations to go to the United States. There were various difficulties for him in leaving Germany. His first task was to obtain permission to take out of the country more than the amount of money permitted by law. The second was to obtain an exemption from the tax which was placed on all those who were emigrating—this alone would be equivalent to about a quarter of his worth. Both of these problems could be solved with comparative ease if he were to be simply on an extended leave.

To support the request for an extension of his leave, Courant argued that his presence in America would in fact be an asset to Germany. He furnished several testimonials to support this contention. Among these was a statement by Ferdinand Springer that it would be in the interest of German scientific publishing for Courant to accept the American invitation.

"When in 1932 efforts were being made in the United States to boycott German scientific literature, Courant—who was then in that country—achieved an improvement in the general situation by negotiating with certain important people," Springer explained. "Difficulties and dangers for German scientific literature in America are going to continue; and for this reason alone it would be desirable that a mediating personality like Professor Courant, who knows the situation on both sides, be in the United States."

The dean of the Göttingen faculty also pointed out that Courant had been responsible for bringing a great deal of American money into Germany and it was important not to sever the connection he represented with the source of that money.

In spite of his personal problems, Courant continued to be concerned about the future of the Mathematics Institute.

Before leaving for America, Weyl had urged Helmut Hasse, at that time a professor at Marburg, to accept a call to Göttingen when and if one came to him. Hasse was an outstanding mathematician, although he was not in the broad mathematical-physical tradition which had flourished at the university. He was known to be politically conservative. Under the circumstances he seemed to Courant, as he had to Weyl, the best possible person to be director of the institute. As soon as Courant received the New York invitation, he wrote Hasse and brought up the subject of Rellich, who was currently a *Privatdozent* in Göttingen.

"If you really go there," Courant told Hasse, "Rellich will be of enormous help to you in the administration and also otherwise."

By Easter, Hasse had received the expected call to Göttingen and was negotiating with the ministry about the terms of his acceptance. At the same time, with a little coaching from Courant, he was trying to support Courant's application for a two-year extension of his leave so that he could accept the New York position without giving up his official position in Göttingen. The minister favored the "emeritization" of Courant rather than another leave and suggested to Hasse that it would be nice if the request for emeritization would come from Courant himself.

Courant had earlier explained to Hasse that there were considerations for him other than financial ones: "My feelings of belonging to Germany and the Mathematics Institute, which after all I created, are decisive. It is clear that even permission for a leave would not make a later return to Göttingen very feasible, for both external and internal reasons; but it is my personal disposition not to turn down the slightest possibility that such a chance might exist."

All during the early summer of 1934, the situation at the Mathematics Institute was chaotic. It was announced that a Nazi party member named Tornier would succeed to the chair of Landau and would act for Hasse as director until the latter officially took over. On his own authority Tornier dismissed Franz Rellich. Courant, shocked and dismayed, communicated this action to Hasse.

"I inform you of this because I think that the apparently premeditated act of not telling you what is being done in your name creates a very bad situation. . . . I hope and wish that things here will soon change so that you can come with pleasure and without hesitation."

But in his heart he was not very hopeful, as a letter he wrote to Carl Still on June 19, 1934, indicates.

Still was trying to promote high quality appointments to the scientific faculty in an effort to rebuild what had been destroyed by the removal of the Jewish professors, and he had asked Courant to help him. Courant felt that he had no choice but to decline. He had come to the belief, he wrote to Still, that the mathematical Göttingen which he and Klein had built up had been irreparably damaged. Germany had not benefited, only countries abroad. America was now trying feverishly to create scientific centers similar to the one that had been in Göttingen.

"The appointment of a few good people cannot put an end to the current anarchy or bring back the regard which the world had for Göttingen."

The day after Courant wrote to Still, the first official extract of Bieberbach's remarks appeared in print. The speech had already become a cause célèbre. Harald Bohr had responded with an answer in a Danish paper based on what Bieberbach angrily called "a ridiculous caricature" of his remarks. Now, on the basis of the published extract by Bieberbach himself, Hardy carefully summarized in *Nature* the ideas which had been put forth and came to "the uncharitable conclusion" that the Berlin professor really believed what he had said. Bieberbach published a strongly worded "Open Letter to Harald Bohr" in the *Jahresbericht* of the German mathematical society. This was done over the opposition of the other two editors, Hasse and Konrad Knopp.

In July, Hasse came to Göttingen to attend to some details of his appointment with the dean of the science faculty. As he walked past the Mathematics Institute, he was greeted by such an unpleasant demonstration on the part of the pro-Nazi mathematics students, led by Teichmüller, that he returned disgustedly to Marburg.

The ministry's response to Courant's request for a leave remained negative. Courant recognized a veiled threat. He requested emeritization—retirement—as of April 1, 1935, but agreed that if the ministry wished for an earlier date for the sake of making new appointments at the Mathematics Institute he would consent. On July 30 Theodor Vahlen wrote to him, "I take this opportunity of sending you my appreciation and special thanks for your valuable academic activities. . . ."

Friedrichs was surprised and impressed that Vahlen had written to Courant.

"I thought it was just a formality," I said.

"Oh no!" he replied. "The Nazis didn't bother with formalities."

From a financial point of view, Courant had come off rather well. His regular salary as a professor was paid to him from the time he returned from Cambridge to the time he left Göttingen. He was excused from the emigration tax in view of the services which it was expected he would perform in the interest of Germany in the United States. He was given permission to take a larger amount of money than was customary out of the country.

With the decision to leave Germany made, he began to delight in the largesse of what everybody called "play money"—the marks still in his bank account, which he could not take to the United States but which he could spend as he wished in Germany.

One morning, before breakfast, Ilse Benfey, a young neighbor, came rushing over to the Courants' house with the news that marriages between Aryans and non-Aryans were to be officially forbidden in Germany. The family of Ilse Benfey's Jewish father had lived in Göttingen for more than two hundred years, but that morning in the summer of 1934 Courant advised her to come with him and his family to America.

"But I do not have the money to do that," objected the young woman, who at an earlier time had given gymnastic instruction to the Courant children.

"Ja, ja. But you should not worry about that," Courant mumbled. His voice trailed off into something about how she could help Nina a little with the children and he would pay her passage.

The elder Benfeys insisted that there was no need for their daughter to leave Germany, but Ilse began to help Nina with the packing. Everything was to be taken, since the shipping expenses would be paid out of the "play money." Heavy furniture that had belonged to the Runges. Two grand pianos, one of them left behind by Hans Lewy, the other purchased by Courant when he was a young officer in Berlin. Innumerable other musical instruments, pictures, books. The collected works of Gauss, of Riemann, of Klein, of Hilbert, of Minkowski. The many volumes, now 43, of the Yellow Series. Mathematical reprints. Letters and diaries. Currency was stuffed into every possible hiding place.

When the moving men arrived, Courant was everywhere, making dry jokes with the men, fussing over the packing of the musical instruments, passing out cigars, tipping everybody generously.

From Göttingen the party traveled by train to Bremen, where they stayed overnight. "What did you do your last night in Germany?" I asked Nina. Her face lit up. "We made music with some friends."

The next morning they sailed for America.

SEVENTEEN

IT WAS on the evening of the tenth day—August 21, 1934—that Courant and his party first saw lights in the distance. Early the next morning they hurried out on deck with the other passengers, several of them also refugees. The day was already sweltering by European standards. Out of a bright morning haze the famous skyscrapers emerged like a range of great mountains. As the *Stuttgart* steamed slowly into New York harbor, a little boy—also a Jewish refugee—greeted "Tante Liberty" with a sweetness and sincerity that touched them all.

At the pier, immediately, they saw Hilde Pick, the cousin who had been Courant's secretary in Göttingen. While they were going through the formalities of customs, several other people appeared to greet them: Franz Hirschland, a German-American industrialist whom Courant had met through the Busemanns on his earlier trip to America; Donald Flanders, the young mathematics professor who had started the negotiations which had led to Courant's new position; a student whom Gregory Breit, a physics professor at NYU, had thoughtfully sent to help them with their luggage. Then suddenly Dolli Schoenberg, one of Landau's daughters, also appeared; and finally, at the barrier, they found Hans Lewy, for the past year an instructor at Brown University. In this situation, as Nina wrote back to Germany, how could they feel that they were strangers?

It had been arranged that, until they found a place to live, Ilse Benfey and the girls would stay in the city in Flanders's apartment while Courant, Nina and the boys would be guests at Hirschland's estate in Rye. They were all packed into two cars—Hirschland's equipped with a uniformed chauffeur—and were driven up through Manhattan, first to lunch and then to Flanders's apartment to leave Ilse and the girls.

Courant had come to the new world resolved to make the best of what he knew would be a modest situation, but he was not prepared for the apartment of the young professor. As he later learned, the self-effacing Flanders and his wife were intellectual individualists who placed their children's education at expensive progressive schools above material possessions and surroundings. They spent their summers on a farm in upstate New York, from which Flanders had come to welcome Courant. During the school year they lived in New York City in what is known as a railroad flat.

As the party climbed flight after flight of stairs, they were increasingly disturbed by what they saw. The flat itself was *schrecklich*—Nina reported —frightful, dreadful. Five rooms, kitchen and bath lay all in a row. Only one room had a real window, and it looked out on the street. How anyone could manage in such a dwelling with two grownups and three children, even for only part of the year, was incomprehensible to the visitors from the big Runge house on Wilhelm-Weber-Strasse. Courant searched in vain for Flanders's study. Outside, some sort of elevated train roared by. He looked at Nina in dismay.

Was this how a professor lived in New York?

The shock of Professor Flanders's home—the lack of a study—was always to be the Courant family's most vivid memory of their arrival in America. In contrast, the Hirschland estate in Rye was everything a home should be in a land of millionaires. A long curving driveway took them through a natural area and then a park-like garden. On the way to the house they passed tennis courts, a lake with boats, stables with riding horses, a garage with yet another car. The house itself was "extremely elegant." In a single room they noted a Cranach, a Corot and an El Greco.

Since Courant was not inclined to describe anything in detail, I looked forward to finding in Nina's letters some account of his first impressions of the university which was to be his new base. But I was disappointed. She describes Flanders: "a blond, somewhat plain-looking young man but exceedingly friendly and modest—certainly no *typical* American"; but she does not devote a word to New York University. In fact, she does not mention her husband's work in her letters until February when one of her correspondents asks, "What about Richard?" From the first she was—as she told me—completely confident that Richard Courant would be as successful in New York as he had been in Göttingen. Her private name for him was always that of the clever cat who impressed the king, outwitted the ogre, and won the hand of the beautiful princess for his young master—"Puss in Boots."

When I tried to draw Courant out about his feelings in August 1934 when he first saw New York University, he said only, "It was very different from what I had been accustomed to in Göttingen."

At that time, although its buildings were scattered all over the city, there were two main locations of the university. One of these, very much like a typical small-town college campus, was situated on the heights of the Bronx overlooking the Harlem River. The other, consisting of a number of buildings without a campus, was at the foot of Fifth Avenue just off Washington Square. This was where all graduate courses met. Courant would have his office at the Heights but would deliver his lectures at the Square, commuting between uptown and downtown by subway. It was still summer vacation, and there were only a few faculty members and students around. One of these was a young man named Morris Kline, who was an instructor working toward his Ph.D. in mathematics. When I talked to Kline in 1971 in his office at the Courant Institute, where he had recently retired as director of the Electromagnetic Division, he told me he remembered very well that first day when Courant had turned up at University Heights. He wanted to know how to go about buying a car— he seemed to be convinced that a person could not exist in America without an automobile—and so Kline found a student to go out and look at cars with him.

In the next couple of weeks the pace and Brobdingnagian scale of American life did much to mitigate for Courant the dismal first impressions of the lot of a professor at NYU. A friend from Germany meeting him inquired, "Ah, Courant, and will you be playing quartets again in this country?" "In this country not quartets," Courant replied. "In this country, octets!" He found many old friends and acquaintances already established in the city and in nearby Princeton, which Nina reported as *sehr göttingisch*. In New York itself the New School of Social Research, which was being referred to as "the university in exile," seemed staffed almost entirely with refugees, as did New York University's Institute of Fine Arts. In fact, Courant and his family found themselves so surrounded by refugees and by German Jews who had come in earlier times that in her letters Nina stopped to describe in detail any "real" American whom they met. She also invariably identified new acquaintances to her correspondents as *arisch* or *judisch*, indicating—to the quarter—the degree of Jewishness. This surprised me—it was so unlike Nina—and then I realized it was the result of what she had so recently experienced in Germany, where a person's future depended on the exact amount of his Jewish blood.

Although Courant continued to follow the activities of the National Socialists in the newspapers, he felt far away from Germany and its problems. And yet—"The situation here is also full of uncertainty and tension," he wrote to Max Born. "But, in the first place, one does not immediately feel and understand it and, in the second place, it is still in seriousness and danger not comparable with that in Europe. I believe, however, that also in America a complete far-reaching transformation of the whole social organization is taking place."

While the children visited from friend to friend, Courant and Nina with the help of Ilse Benfey tackled the problems of settling a family in a strange country. They had been in the new world just two weeks when they rented a house, and Nina was able to write to her mother, "We now have a real address, 142 Calton Road, New Rochelle, USA." It was to be Courant's home address for the rest of his life, and the house where Nina still lives with Ilse Benfey, now retired after a long and active career as a social worker.

New Rochelle, located on Long Island Sound, northeast of New York City, is some twenty-five miles from Washington Square, where Courant was to teach. Friends had recommended the town for its good public schools. Nina, who had some Huguenot blood, was pleased to note that its name and its architecture in many sections memorialized the French Protestants who had fled their homes in Europe to escape persecution for their religion. The principal monument, a few blocks from the Courants' new home, was a statue of Jacob Leisler (born in Frankfurt am Main), who made it possible for the Huguenots to settle in New Rochelle.

The house which the Courants had rented was a big, comfortable, but very modest structure standing upon a knoll with other similar houses. They found it one of the curious things about America that people often lived in such wooden houses all year around, not just in the summer, and also that there were no fences between their garden and the gardens of their neighbors. A pleasant tree-lined street led down to the high school, in front of which there was a little park and a small lake. There was a garage, surprisingly "right on the street"; and within a week after they had moved in, Courant had negotiated the purchase of a secondhand Chevrolet through a cousin who was an insurance agent—one of the four Courants of his generation already in America in addition to Hilde Pick and himself.

Hilde Pick agreed to act again as his secretary and came out to New Rochelle to live with the family as she had in Göttingen. Ilse Benfey took over the management of the household and the children. Nina began immediately to "make music." The change in the family's situation seemed hardest on Ernst and Gertrud, the two older children; but every evening as 10-year-old Hans lay in bed, his mother heard him singing *Deutschland, Deutschland über Alles* to himself, "in the softest tones of which he is capable." Six-year-old Lori marveled that no one in school asked her if she was Jewish.

What was it like, I asked Courant, to leave his home and country and start over at forty-six in a situation where he was again financially and professionally at almost zero.

"It was hard," he said simply. "There was such a little bit of money. I felt myself responsible for many people. There was really nothing scientifically at NYU. And then of course"—his voice trailed off as it usually did—"I was so attached to Göttingen."

When I asked the same question of Nina, she replied, "Of course it was something we could never have brought ourselves to do on our own, but when it was forced upon us and there was nothing else we could do, it was wonderful—like being young all over again!"

I have been told that, on the occasion of their fiftieth wedding anniversary, when Courant in a little speech described Nina as "a heroine" who had courageously left family and friends to follow her husband to a strange land and a precarious position, she had stood up at the table and, rapping on her glass, had asked to say a few words herself. It had not been the way Richard had described it at all. In fact it had been rather a relief to her to leave the little university town of Göttingen where she had grown up as one of the four daughters of Professor Runge and where she and her family had lived, after her father's death, in the old house that had belonged to her parents. She had been no heroine, she insisted. She had been happy and eager to leave the old world for the new!

Watching Courant's face while Nina spoke, many of the guests had the feeling that he was displeased. For him it had been different.

After he arrived in America, he continued to seek news of the Göttingen institute from European friends outside Germany. In September, he received a firsthand report from the English mathematician Harold Davenport, who had recently visited Hasse.

In the late summer of 1934, at the urging of other German mathematicians, who were disturbed by what had happened to mathematics in Göttingen, Hasse had agreed to go back to that university; but his life was still being made miserable by Tornier and the pro-Nazi students led by Teichmüller.

Davenport's letter also brought news of the annual meeting of the *Deutsche Mathematiker Vereinigung*, at which Tornier had appeared in the company of a storm trooper in civilian dress (later asked to leave). Although the election of Blaschke as president had been considered a victory for the moderate forces, the members of the DMV had passed a resolution censuring Harald Bohr "most sharply" for his attack on Bieberbach "to the extent that it was an attack upon the new German state and upon National Socialism." They had deplored the fact that Bieberbach had published his reply to Bohr over the opposition of the other editors, but they had formally recognized that he had been motivated by his concern "for the interests of the Third Reich."

Courant was incensed by the affront to Bohr. He also felt an urge to write a few lines of sympathy to Hasse, "but I am afraid such a correspondence observed by the Tornier guards may do him more harm than give him comfort."

The students Courant faced at Washington Square were very different from those he had known in Göttingen. He found them "not ungifted but extraordinarily poorly prepared." Almost all were Jewish, the sons and daughters of immigrants from eastern Europe. As undergraduates they had usually attended the City College of New York, since it charged no tuition. After graduation they were able to continue their education with a class or two at NYU only because that university regularly offered its graduate courses in the late afternoon and evening.

Flexner described these young first-generation Americans from the nation's largest city as "a great reservoir of talent," and Courant now made this phrase of Flexner's his own. In the coming years he was to use it over and over. The metaphor touched a responsive belief in him.

"There are many outstanding people, potentially, everywhere at all times," he told me, "but of course the conditions are not always conducive to their development."

"That is something you really believe, isn't it?"

"Very much."

The role Abraham Flexner played in helping Courant during his first years at NYU is not generally known, and it was in fact so minor in relation to Flexner's many other activities that he does not mention it in his autobiography. His correspondence indicates, however, a sincere interest in the project and in Courant, whom he saw as an "idealistic, energetic, and unselfish" man. During the years from 1934 to 1939, when Flexner retired as director, the Institute for Advanced Study cooperated with Courant in various practical ways. Flexner personally also gave him introductions to wealthy New Yorkers who might be interested in the improvement of the graduate mathematics program at New York University.

"How *was* the program being offered in mathematics at NYU when Courant came?" I asked Morris Kline.

"No better than mediocre, maybe even a little worse," he replied promptly. "There were only a few people who could lead doctoral candidates, since not all members of the faculty had doctor's degrees themselves. I think that when Courant came I would have shifted over to him if it would not have meant going into an entirely new area of mathematics. I was interested in topology then—like most young American mathematicians of the time."

Kline's evaluation of NYU's graduate mathematics program in 1934 is borne out by a study which was being concluded that same year by R. G. D. Richardson in the hope that, as he wrote, "those concerned with the strategy of promoting mathematical thought and achievement in America can find in [it] several signposts for their future guidance."

As one criterion for the excellence of the program at a given university, Richardson took the number of Ph.D.'s which had been awarded. (Another was the amount of work produced, as measured by published pages.) In the course of his survey, he discovered that since 1862, when Yale had awarded the first doctor's degree in mathematics, more than one-sixth of the total of 1286 degrees had been conferred by the University of Chicago. Six universities (Chicago, Cornell, Harvard, Illinois, Johns Hopkins and Yale) had been responsible for more than half. In the decade immediately preceding (1924–34), only 29 universities had awarded five or more doctor's degrees and could thus be considered an important factor in the mathematical education of America.

New York University was not one of these.

Harry Woodburn Chase, the chancellor at NYU since 1933, was definitely interested in improving the graduate mathematics program at the university. Unlike Flexner, whose father had been an itinerant peddler of hats when he first came to America, Chase had a background completely different from that of most of the NYU students. He came from an old American family, had grown up in a small New England town, and had attended Dartmouth College. He had taken Courant with the assurance that Flexner, who had earlier refused the chancellorship, intended to maintain a practical interest in mathematical developments at NYU.

"The organization and the inner workings of the giant university are very complicated, and it will be a long time before I understand them completely," Courant wrote to Born after meeting Chase. "It is clear, however, that—in principle—a really rewarding project offers itself here. I am not quite sure how it should be carried out. Also it is by no means certain that I shall have the opportunity."

The ties with Göttingen which he had struggled so hard to maintain during the past year began to be severed almost immediately. Notification came from Germany that his emeritization, which had been set for April 1 of the following year, had been moved up to October 1. This formal conclusion of a career as a German professor—under normal circumstances—would not have occurred until 1956. On the first day of October 1934, Courant wrote drily to Franck, "Today I celebrated my sixty-eighth birthday."

Before the end of the year another tie was to be severed. Protesting the censure of Harald Bohr "for remarks which he had made as a private person," Courant resigned from the *Deutsche Mathematiker Vereinigung*, which had been founded in happier days by Klein and Hilbert and others.

Franck, concerned about his friend's future in a new country where people might not understand and appreciate his virtues, cautioned him:

"Please, dear Courant, prepare your lectures well and try not to organize right at first!"

Courant intended to follow this advice and, as he put it, "to regard present conditions as passively as possible and to try to give good lectures from which people can really get something."

He hired a man in the speech department to tutor him in English and carried in his pocket a small notebook in which he was constantly writing down colloquialisms he came across in conversations or in books. He dictated carefully prepared lectures and then translated them from German into what he hoped was "good English." Although he also dictated innumerable letters in German to friends still abroad and to his other German friends now in America, he tried his hand at composing letters in English whenever an opportunity offered. Even the American Cancer Society, New Rochelle branch, received a long apologetic account of his financial situation as a refugee professor along with his check for $5.00.

He had done no real mathematical work since the Courant-Friedrichs-Lewy paper. He knew he should begin to publish again, but he found himself unable to do mathematics or to work on the still unfinished second volume of Courant-Hilbert. He could not accustom himself to the quiet, the lack of people around him, the silent telephone. Sometimes, in his little office at the Heights, he would lift the receiver from its hook and listen in the hope that there had been some mistake and the phone had simply failed to ring.

In spite of his resolution to follow the admired Franck's advice, he was eager to be "genuinely helpful" at NYU. He saw many things that should be done. The worst aspect of the situation was the lack of an accessible mathematical library, which made it virtually impossible for students to do independent mathematical work. He was also soon aware of a need for financial support on the part of some students if they were going to be able to concentrate on their education. He began to talk to Hirschland and other wealthy German-Americans with whom he had become acquainted about "the reservoir of talent" at NYU and the desperate need for a *Lesezimmer,* for *Assistentstellen*—the German words came more easily. Some of these men gave him small sums in dollars. Some also offered blocked marks—the "play money"—which they had in German banks and were unable to spend outside Germany.

All during his first year he mulled over the problem of the frustratingly blocked funds. These could be used to purchase German books and journals only if they were matched with American dollars. He worked out a plan for cooperation between NYU and other universities but abandoned it as too complicated. He also wrote to the American Library Association suggesting that it might be feasible for the organized American libraries to obtain from the German authorities the right to pay a considerable part of their orders in blocked marks and so reduce the cost considerably. This suggestion did in fact later bring results, but it was considered more diplomatic by the association not to give Courant credit for it.

Unobtrusively, so he thought, he tried to establish contacts which might turn into future sources of funds. The semester had hardly begun when he hunted up Warren Weaver, the director of the Division of Natural Sciences of the Rockefeller Foundation. He also recalled himself to Henry Goldman, a wealthy German-American Jew, a friend of Born and Franck who had at one time contributed some money—"not very much"—to the physics institute in Göttingen. (Rich people, Courant always said, have no real conception of money.) With his introductions from Flexner, he met various wealthy Jews in the city.

Ultimately, he knew, it was going to be *most important* to upgrade the faculty at NYU in mathematics and in the related sciences. Breit was no longer in the physics department; and except for the selfless Flanders, whom Courant was always to refer to as "a saint," he felt very much alone in his new position. A few months after his arrival, he heard that the Rockefeller Foundation was willing to sponsor Siegel's coming to the United States. He tried to arouse interest on the part of the NYU administration in providing a place for the German mathematician; but instead Siegel received an appointment at the Institute for Advanced Study.

Courant went eagerly down to the pier to meet him when he arrived in January 1935 and transported him out to New Rochelle to stay until he had to report in Princeton.

"The most important news," Nina wrote to her family in Germany, "is the arrival of Siegel, one of Richard's first assistants, a postman's son, big, very Aryan. . . . In Richard's opinion, a mathematician of Hilbert's stature."

Less than a month later, Siegel came back to New York to deliver a lecture at Washington Square to Courant's students and invited guests. This lecture was only one of a number of lectures by well-known mathematicians which Courant arranged during his second semester at NYU. Although most of these were set up by him to give his students insights into areas of mathematics outside his own specialties, the subject of one lecture that semester was in the area which most appealed to him and was, in the course of the coming year, to bring him back at last to mathematics.

This was the lecture in March 1935 by Jesse Douglas, a mathematician at MIT who had attracted worldwide attention by solving Plateau's problem, one of the oldest and most famous problems in the calculus of variations.

Courant's invitation to Douglas to speak at NYU was to a certain extent an olive branch. Since coming to America, he had made a determined effort to remove what he called "dissonances" in his relations with American mathematicians who had earlier left Göttingen feeling somewhat offended by their treatment there. Douglas was one of these. Arriving in the year 1929–30 at the end of a European tour as a National Research Fellow, he had proposed to talk upon his not yet published work on Plateau's problem at the weekly meeting of the *mathematische Gesellschaft*. The problem had been around for a long time. Many outstanding German mathematicians, including Riemann himself, had worked on it. The members of the *Gesellschaft* simply did not believe that an American had solved it. As it happened, Douglas's solution, which was highly original, did not appear to be in completely rigorous form. When he finished his presentation, some of the members of the *Gesellschaft* took him severely to task on almost every detail of his proof. He left Göttingen deeply offended but determined to show the people there that his argument had been correct. When he finally did publish his work in 1931, he laid out the chain of reasoning in an unassailably rigorous fashion.

Remembering this past history, Courant did his best to make Douglas's lecture at NYU an event. He sent out notices to other colleges and universities in the area and saw to it that Douglas's former teachers at Columbia and CCNY received personal invitations. In view of the level of the students, however, he suggested that Douglas try to keep his remarks "as elementary as possible." It was a long way from a meeting of the *Gesellschaft*, but afterwards there was coffee and discussion as in the old days. Courant asked Douglas to send him enough reprints of the famous work so that every member of his seminar the following year could have one.

After I learned some of the details of Courant's first year in America, I became curious what idea of the future had been in his mind when he came here. I asked him if from the beginning he had hoped to build up another institute at NYU like the one he had had to leave behind in Göttingen.

"Well, that is difficult to say," he replied. "Everything is so different in the perspective of so many years. When I came here I felt some kind of— I don't like such words—but some kind of *patriotic* urge to do something for this country. I was deeply impressed by America and what was being done during the first years of the Roosevelt administration. I was very enthusiastic. I felt very loyal and thought very much about what was needed. The best I could imagine that I might be able to do was to bring my experience in Göttingen to bear upon the situation here."

EIGHTEEN

COURANT was not kept long in suspense as to who was going to have the opportunity to develop "the really rewarding project" which he saw as existing at New York University. In June 1935, at the end of his first academic year in America, he received a note from Chancellor Chase informing him that when his temporary appointment expired the following June the university was prepared to offer him a permanent place. In the note Chase stated he believed that, in conjunction with the Institute for Advanced Study, there was a real opportunity for Courant to develop at NYU a strong graduate department of mathematics in ways that would be interesting to him.

"I am saying this now," he wrote, "because I want you to feel whatever sense of security that it may bring."

It was the moment when, in Courant's opinion, "a continuous work" should begin. There was still no space at Washington Square for an office or a library, but there were some encouraging developments. Goldman had offered to give $500 for the library if Courant could raise another $2500 on his own. The ever helpful Flexner had agreed that members of his institute at Princeton could give individual lectures and even full courses at NYU during the coming year if Courant could guarantee their train fare. George Blumenthal, a banker who had sat beside Courant at an official university dinner, had donated $1500 for fellowships in the fields of Courant's mathematical interests.

The money from Blumenthal was the first Courant had had the opportunity to distribute in America, and he took up the task with enthusiasm. One fellowship went to Irving Ritter, German-born but an American citizen, who had been a graduate student at NYU during the preceding year. He was older than the other students, married, and working as a nightwatchman. He later became a professor at University Heights and is now retired. Another fellowship went to Max Shiffman, the outstanding mathematics student in the current graduating class at the City College of New York—now at the California State University at Hayward.

Having awarded one fellowship to a German-born American and the other to an American who was Jewish, Courant concluded—and others agreed with him—that the third fellowship should go to some "really good 100 per cent American." At the beginning of 1935 he sent out a number of letters to other mathematicians asking for recommendations. By August, when no suitable candidate had turned up, he suggested to Flanders that for the third fellowship they push Rudolf Lüneburg, one of his students from Göttingen, who had left Germany very early, even though he was not Jewish. At the last moment J. D. Tamarkin of Brown University, himself Jewish and a Russian immigrant of an earlier day, came up with the suggestion of a young man with the completely satisfactory name of Tom Confort. Courant managed to find some other way to support Lüneburg, who after a year at NYU went into industry and made outstanding contributions to optics. I asked several people what happened to Confort, but nobody knew.

In spite of Courant's efforts in connection with the development of a graduate mathematics program at NYU and his active concern about placing friends and former students from Germany, his top priority in the summer of 1935 was still the completion of the second volume of Courant-Hilbert. Already many of his letters in English had assumed a format which was to become standard with him: "I have a very bad conscience that I have not answered your letter . . . which I must admit slipped between some other mail and was lost for a time . . . but. . . ." During the first year in America the "but" introduced the statement that he was "under very much pressure" to finish Courant-Hilbert II. In the months since his arrival he had made almost no progress; however, in August 1935—at the end of the German summer semester—he expected Friedrichs for a visit. He was again optimistically certain that with Friedrichs's help he would at last be able to finish the book.

Friedrichs had a reason for coming to the United States other than helping Courant. He was in love with a Jewish girl whom he had met just four days after Hitler had become chancellor. Although the Nuremberg laws prohibiting marriage between Aryans and non-Aryans had not yet been enacted, all official National Socialist newspapers were regularly publishing the names of non-Jews who were alleged to have had relations with Jews. In a number of towns individuals had been sent to concentration camps as a result of such charges. To protect their family and friends as well as themselves, Friedrichs and Nellie Bruell had been able to meet only clandestinely during the past year. They were considering whether they could emigrate.

Courant was delighted to have his former student and assistant in New York, to talk to him, to show him around, to introduce him to the important people he had become acquainted with, to explain America to him. He produced several new odds and ends of manuscript and the first two chapters, which had already been written when Friedrichs had first started to help with the book ten years before. "Now we will finish!" Courant insisted. But nothing much was accomplished that summer either, except that the first two chapters were finally "really written."

The day before Friedrichs was to return to Germany, he and Courant went for a long walk. Friedrichs announced that he had decided he could be happy living and working in the United States. Courant advised him to emigrate immediately—the fact that there was as yet no academic position available for him should not stop him.

"The traditional way to come to America," Courant mumbled, "is as a dishwasher."

Friedrichs answered mildly that he was not yet ready to exchange mathematics for dishwashing.

By the time Friedrichs left to return to Germany, the American semester had already begun. He remembers that Courant was very excited about the seminar he was conducting on conformal mapping, minimal surfaces, and Plateau's problem. The students were interested, intelligent, and willing to work; and one of them—young Shiffman—was in Courant's opinion exceptionally gifted. Plateau's problem was attractive to Courant; and even before Friedrichs left, he was showing signs of interest in working on it himself.

"But he didn't *want* to get involved in that," Friedrichs told me. "He *wanted* to finish the second volume of Courant-Hilbert."

In addition to trying to finish Courant-Hilbert II, to lecturing and conducting the seminar, to searching for sources of support for his enterprise at NYU, Courant was constantly concerned with the personal and professional problems arising out of National Socialist policies toward German Jews. Both his parents were now dead. His remaining brother, Fritz, and his family were in Italy, where Mussolini had not yet adopted Hitler's anti-semitic policies. Courant's interest and activity usually did not extend to his Courant cousins and aunts and uncles still in Germany, but he *was* concerned about his friends. He mounted a massive campaign to get a visa for Ilse Benfey, who had come originally as a visitor. Helping her was complicated by the fact that she was not a relative. Courant, however, managed to get Flexner, Flanders, Hirschland, Goldman, and a number of other important people to write to the effect—as Nina described it—"that Richard is a really splendid fellow and in a position to give an affidavit of support for Ilse, who also because of her own capabilities is not likely to become 'a public charge.' " Eventually, under the force of all this ammunition and Ilse's own frequent calls, the American consul in Hamburg yielded, so flustered that he signed Ilse's last name instead of his own on her visa.

Letters asking for help and advice came "by the dozens" from mathematicians in Germany. No one who had known Courant there could believe that he was not so knowledgeable and effective in New York as he had been in Göttingen. He worried the most about the future of former students. Since March he had fretted over the case of Fritz John, whose grant from the Academic Assistance Council in England was going to expire in June, leaving him and his ailing young wife virtually destitute. Then, in the fall of 1935, Fritz John received an unexpected appointment at the University of Kentucky. Courant rejoiced. It was an exceptionally good position for a man so young and with so little experience!

Back in Germany, in Göttingen, things seemed to be settling down.

When the students had placarded the Mathematics Institute with signs proclaiming that Hasse, who was the treasurer of the *Deutsche Mathematiker-Vereinigung*, "permitted Jews in the mathematical society," Hasse had gone to Theodor Vahlen in Berlin and had demanded that something be done about the situation in Göttingen.

He told me when I talked to him in San Diego:

"I tried to appeal to his mathematical soul—not his political feeling. Then he said to me—I can see him sitting there, a broken man—'Yes, I see all that, but I can't help. I am a weak man.' "

Tornier, nevertheless, was shortly transferred to Berlin. There he embarrassed the mathematics faculty by being pictured in the newspaper walking on a fashionable boulevard with a notorious prostitute on his arm and a tame tortoise on a leash.

In the fall of 1935 Davenport reported to Courant after another visit to Hasse:

"The number of students is extremely small. . . . Witt is Hasse's personal assistant, and he certainly is a good mathematician. . . . I gather that Teichmüller, who was the ring leader of the opposition to Hasse among the students, has become reconciled to him. Hasse thinks he is quite a good mathematician, but I am unable to judge. . . . Kaluza, of Kiel, has been appointed to one of the chairs at G. (I think yours) and is starting this term. . . . Hasse is trying to get Deuring to G. . . . , but all attempts meet with great delay in the ministry of education. In fact, the university business as a whole seems to be greatly neglected. . . ."

Outside every village and town there were notices posted: *Juden sind nicht erwünscht*—"Jews not wanted." There was also a box in which citizens could deposit accusations against those who had any association with Jews.

"How decent people can tolerate such things is absolutely incomprehensible to me," Davenport wrote. "If it were not for my friendship and mathematical interests in common with Hasse, I would not dream of going to Germany. . . . In fact, one has the feeling that the greater part of the population is mad—one of the characteristic features of such madness being that victims are normal and (frequently) delightful people—on all subjects but two or three."

It is hard to understand how at this time Courant could have considered a trip to Germany, even to take advantage of the blocked marks which friends had contributed for mathematics at NYU; but in his files there is a letter to Alwin Walther, a former assistant and a professor in Darmstadt, describing plans for such a trip in November or December 1935. Across the letter the word *Nicht* has been scrawled with a red pencil.

That same November for which he had planned his trip, Courant received news of the death of Julius Stenzel, the friend and mentor of his youthful days in Breslau.

"I was truly touched," Nina wrote her mother, "by the letter Richard wrote to Mrs. Stenzel, in which he said that the friendship with Julius had been a high point of his life."

Although Stenzel had not been Jewish, he had been removed in 1933 from his position as professor of philosophy at Kiel because of earlier disciplinary acts against pro-Nazi students while he was rector of the university. He had then later been sent to the university at Halle. His widow was the former Bertha Mugdan, whom Courant had once tutored in mathematics and physics. She was Jewish according to Nazi racial laws, although she had been a convert to Christianity since before her marriage to Stenzel. In his letter of condolence, Courant urged that she and her four children emigrate to the United States immediately. He offered to furnish the necessary affidavits guaranteeing their support.

Siegel had returned to Frankfurt after he had learned that Werner Weber, who had forced Landau out of the lecture hall, had been sent to that university. He had felt, according to Courant, that it was his duty to go back and fight for his colleagues, especially Max Dehn and Ernst Hellinger. But by the time Siegel arrived in Frankfurt, both Dehn and Hellinger had been removed from their positions.

In conjunction with Oswald Veblen, who spearheaded the American effort to find places for refugee mathematicians, and Harald Bohr, who was the European contact, Courant began to think also about getting friends who were not Jewish out of Germany. He was particularly worried about Artin, whose wife was half Jewish.

It was against this background of concern about his friends that Courant finally gave way to the attractions of the Problem of Plateau. The problem—which gets its name from that of a nineteenth century Belgian physicist who conducted physical experiments in connection with it—is very deep mathematically. It can, nevertheless, be stated in a way that it is intuitively clear even to people who are not mathematicians. In its simplest form it is *to find the surface of least area that spans a given closed contour in space.* In spite of this easily grasped formulation, it is far from obvious how to determine the solution analytically; and, except for very simple contours, it is not even clear, intuitively, how the surface should look.

Courant had been very much impressed by Douglas's achievement in finding the first complete solution of Plateau's problem for single and double contours. The American had ingeniously employed a peculiar minimum problem, to which he had been led by the transformation of yet another minimum problem related in turn to the minimum problem of Dirichlet. It was a highly original piece of work, and in Courant's opinion Douglas had justly earned credit for it all over the world. Much as he was impressed by Douglas's achievement, however, he found the method the American had used unappealing. It was not, in his view, nearly so simple and direct as the nature of the problem deserved. Douglas's approach seemed to Courant "roundabout" just as, more than a decade before, Weyl's approach to the problem of the Lorentz conjecture had seemed "roundabout."

"Courant would never tackle a problem just in order to solve it," Friedrichs explained to me. "The tools used to solve the problem had to appeal to him too. He could never separate them from the problem. They were always part of the deal. If the problem was beautiful, the tools had to be beautiful too. Otherwise, there was no appeal at all."

Courant was aware that mathematicians had long recognized the connection of Plateau's problem with harmonic functions and the Dirichlet problem. In his seminar in the fall of 1935, with his gift—reminiscent of Klein—for weaving connections in mathematics, he glimpsed a way in which the minimum problem related to that of Dirichlet could be employed in a straightforward manner for the solution of Plateau's problem. Again he reached for his most effective tool, the lemma implicit in his dissertation of 1910 and later explicitly stated in 1914. It was the same lemma which he had also used so effectively in the work on finite difference equations. By February 1936—a little less than a year after Douglas's lecture at NYU—he was able to write enthusiastically to Siegel that he had been able "really importantly" to simplify and extend the solution of Plateau's problem so that the whole affair—in his words—"is now no more difficult than the fundamental theorem of conformal mapping."

At home he played a great many Bach fugues.

"There is nothing but good news to report from here," Nina wrote. "Richard has done more scientific work than in a long time—including in the nice Göttingen time—and he is very happy because he has found something."

Although Nina was a mathematician's daughter and had more than an ordinary talent for the subject, she had long ago given up trying to understand her husband's mathematics.

"To be able to work with one's husband in his own subject must be really wonderful!" she wrote to her mother. "Had I been more energetic (and poorer!) I would have learned to do Richard's secretarial work, but I am afraid that I would have done it wretchedly. Still, this working together of husband and wife is very modern. I think it would be splendid!"

"My relationship with mathematics is rather like that of a small child making its first discoveries and *playing* with things," she told me.

"It enjoys with fresh wonder and with huge joy that a marble will roll or that the pieces of a puzzle fit together. That is how I felt when I first learned to have letters represent numbers or geometrical units in formulas and equations. And then, having seen some mathematical facts, I was not quite satisfied with that, but I wanted to know: why is this so, what is behind it, what follows from it? However, I immediately forgot my results and remembered only the exhilaration they had given me. Being a lazy person, I was never tempted to work my way into higher mathematics.

"What I really loved was to be present when Richard was having mathematical conversations with Friedrichs or Fritz John. Not understanding a word they said (nor trying to), I could feel their eagerness to learn from each other—there was a kind of peaceful agreement and mutual respect between them which I never saw between people in other fields."

In his professional and personal activities up to the time of his work on Plateau's problem, Courant had made a considerable effort to efface himself. He was well aware of the native American's distaste for anything that could be construed as "meddling" by foreigners, especially when the foreigners were also Jewish. He had also observed how recent immigrants seemed especially threatening to immigrants of an earlier day. During his first years at NYU he almost always put forth his proposals through Flanders, "a genuine 100 per cent American," whose ancestors had actually been in America before the *Mayflower* arrived.

In spite of these efforts "not to offend," he received a disturbing letter from Franck in the winter of 1935–36. From people "irreproachably loyal," Franck had heard that in Princeton and New Haven "and perhaps other places as well" there was a feeling that Courant was being too forward for a foreigner.

"I know that your goodness and helpfulness cause you to be concerned about people," Franck assured him, "and I myself have often enough called upon you for help in the case of my own children, but in spite of that I beg you most sincerely to take this report to heart and hold yourself 100 per cent back."

It seemed to Franck that hostility to foreigners was growing—he and others often felt there was more anti-semitism in the United States than there had been in pre-Hitler Germany—and that Courant should not in the future write to places other than New York University unless he had been asked expressly to do so.

"If you don't follow this advice, you may be able to help a couple of people that otherwise you couldn't help, but the damage you will do to yourself and your friends will be greater in total."

Perhaps, Franck conceded, a similar feeling against himself existed.

"But by accident I have heard these rumors about you, and they must serve for both of us as a warning to be enormously careful. We dare not forget that we already once before deluded ourselves about the firmness of the ground on which we stood."

Carefully examining his actions since coming to America, Courant came to the conclusion that there was really no *objective* basis for the criticism Franck had reported.

"I can't believe that they are 'the' mathematicians at Princeton or at Yale. In Princeton I am with Weyl, Veblen, and Neumann on *absolutely* friendly and open terms. Likewise with Lefschetz. With others I have never talked about questions concerning German emigrés. But there is naturally in Princeton another group with whom I have much less contact. These people, who are in opposition to Veblen's group, were seemingly against Siegel's appointment and may have connected me with that. I have never been at Yale and know only Hille closely. He has always been a candid friend and has often spoken with me in general about the hostility toward foreigners."

It seemed to him that his conscience was clear.

"But I understand completely that I must shun the slightest appearance [of meddling] and that my propensity for harmless utterances that signify nothing is dangerous. I will, therefore, seek to be still more cautious. . . ."

The resolution was scarcely made when Courant found himself in an unpleasant position with an American mathematician as a result of his work on Plateau's problem.

Courant's approach, in addition to giving solutions for problems of single and double contours, as Douglas's had, could more readily be carried over to higher cases in which the boundary consists of any number of distinct contours. When Courant's results establishing this were published, Douglas insisted that the same thing could be done by his method, that it was in fact implicit in his work. He pointed out a paper, which he had already submitted, in which he had applied his method to the case of several contours.

The situation resulted in some bad feeling. The American was in a fragile mental condition at the time and was shortly to be institutionalized for a period. Between 1936 and 1938 he had no academic position.

In 1937 Courant rather sweepingly gave Douglas full credit for his method for the general case and stated that his publications did not contest Douglas's claim of priority.

During the first half of 1936, Courant published three papers on Plateau's problem. It was the beginning of one of the great productive periods of mathematical creation in his career—when he was almost fifty —and the first real mathematical achievement since the work on finite differences.

"You see, in the last years in Göttingen I had lost my contact with mathematics a little bit," he explained to me. "But then here in this country I had to get on my own feet, and I was very happy that I started doing mathematics again. That was very satisfactory to find that I had not lost my competence and was able to do something."

A few months before his death, I asked him what he recalled as the most mathematically exciting event of his life. He placed the return to mathematics in the 1930's in the same class with his youthful contact with Hilbert.

"So in a way I was really grateful for that. I thought I was the beneficiary of Hitler."

He continued to make little progress on the second volume of Courant-Hilbert, but he had reason to be pleased with developments at NYU. The number of graduate students had grown as his presence in the city had become known. Many who attended his classes were men already established in fields that involved the applications of mathematics. Immigrating scientists and foreign visitors stopped to attend his lectures as they passed through New York. The arrangement with the Institute for Advanced Study had resulted in a number of exciting individual lectures and a broadening of the graduate offerings. Also, and most important, Chancellor Chase had promised that in September 1936 space for a library and an office for Courant would be made available at Washington Square.

It was at this time, during his second spring in America, that Courant first began to talk to people with access to large amounts of money about his plans for a graduate center of mathematics at New York University. One of these was Dr. Frederick Keppel, to whom he had been introduced by Flexner and through whom he hoped to interest the Carnegie Foundation. Keppel was friendly, but he told Courant that *just* building up a mathematics department at New York University was not going to appeal to the Carnegie people.

"Therefore, he suggested that one should prepare the attack a little more on tactical lines. . . ," Courant explained to Flexner, "how we are attempting, so to speak, a new experiment in graduate education . . . and reasons why just New York University and why mathematics. Fortunately what we have in mind can, without any artificial effort, be presented in such a way."

"I wanted to do something like what Flexner had presented in his book on universities, which I had read and which had very much impressed me," Courant explained to me. "This was not done at his own Institute for Advanced Study, because there was no teaching there and no contact with the applications."

Interwoven with Flexner's ideas was Courant's own personal conception of Göttingen as a place where no distinction was made between mathematics and its applications and where advanced students and faculty were like a family. The "twist" would be to transplant this essentially elitist institution with its noble scientific tradition to a mediocre, business-oriented American university with no tradition at all and a student body composed largely of the sons and daughters of working-class Jewish immigrants.

In June 1936 Courant received the permanent appointment that Chase had promised and an increase in salary of $1000 a year. Chase also announced in his annual address to the Board of Trustees the intention of the university to establish "a graduate center for mathematics" with Professor Courant as its director.

Space for the new center was obtained in a building recently purchased on the north side of Washington Square. It was only three small rooms upstairs in a house that had been on the Square since the days of Henry James, but Courant was happy. Mathematics at NYU now had an address. Remembering from his Göttingen days the power of a name on a letterhead, before he left for the International Mathematical Congress being held that summer in Oslo, he ordered stationery printed:

NEW YORK UNIVERSITY
The Graduate Center for Mathematics

NINETEEN

THE NEW Graduate Center for Mathematics had a faculty of one—Courant himself—but already, as he set off for Europe in the summer of 1936, the combination of circumstances which would provide him with two young colleagues perfectly cast for his purposes had been put into motion.

During the preceding year he had heard from his Göttingen friend Heinz Hopf, by that time a professor in Zurich, about a young American engineer, an assistant professor of mechanics at the Carnegie Institute of Technology, who had taken his Ph.D. under Hopf with a thesis on a topic in differential geometry. The combination was rare: an American, an engineer, an interest in pure mathematics. Courant had immediately written to J. J. Stoker: "In case you should come to New York sometime at the end of this or at the beginning of the next academic year, I should be glad if you could get in touch."

That same spring, Courant had been trying to arrange some sort of position that would enable Friedrichs to come again to New York. Although he had been successful in obtaining only enough money to defray the younger man's living expenses, he had dispatched a formal invitation in English for a few lectures at NYU. To his surprise Friedrichs had wired back in German, "Unfortunately I am not in a position to accept your friendly invitation." Mystified by the refusal, Courant was looking forward to seeing Friedrichs at the International Congress in Oslo and receiving an explanation.

When Courant arrived in Copenhagen, he found Friedrichs already there. The young man explained that the fact that Courant's invitation had been in English had brought a rebuke from the official in the ministry to whom he had had to apply for a leave.

"I can find no explanation," this man had written angrily to Braunschweig, "why Professor Friedrichs has not already declined the invitation from Professor Courant, one of the Jewish emigrés who, although he knows German, writes in English. I want a copy of Professor Friedrichs's refusal of the New York University invitation sent to me immediately."

Under the circumstances Friedrichs had thought it would be more discreet to meet Courant in Copenhagen rather than at the congress in Oslo, where their meeting would be observed. He told Courant that he was now prepared to emigrate if Courant could arrange a professional position of any sort for him. Courant was not optimistic. There were so many Jewish mathematicians, who *had* to emigrate, and so very few positions. He mentioned again that the traditional way to come to America was as a dishwasher, but Friedrichs again declined that alternative.

The congress at Oslo was marred for Courant, Weyl, and other German refugees to the United States by the absence of a number of colleagues who had remained in Germany. On the American side there was also a notable absence. Jesse Douglas and L.V. Ahlfors had been chosen as the first recipients of Fields Medals, but Douglas had not been able to afford the trip to Oslo. His award had to be accepted for him by Norbert Wiener.

Although Courant had given Nina "a thousand reasons" why it would not be wise for her to visit Germany at this time, he himself planned to pass through that country on his way to Carlsbad, a favorite spa in Czechoslovakia. He found it strange and a little bit frightening to enter Germany again; but it was the summer that the Olympic Games were being held in Berlin and Hitler was showing off to the world the accomplishments of three years of National Socialism. And so, as he said, "There was nothing difficult."

In Hamburg he saw Artin, who had not been given leave to attend the congress in Oslo. Artin now wanted to visit the United States with his family and see how he liked it before deciding whether to emigrate.

From Hamburg, Courant went to Berlin, where he made arrangements for blocked marks in his account to be placed at the disposel of his friend Stenzel's daughter Anna so that she could emigrate. He also saw Springer, who was apparently still doing business very much as usual. Although the Nazis had forced a party member upon him as a "partner" in place of his cousin Julius, who had been found to be more than half Jewish, there had been no further interference in the activities of the firm. Courant was still editor-in-chief of the Yellow Series. It was still planned that Springer would publish the second volume of Courant-Hilbert. There were still Jewish names on the title pages of Springer journals. An obituary of Emmy Noether, who had died in the United States in the spring of 1935, had appeared in the *Mathematische Annalen*.

I asked Courant if he had gone to Göttingen on this trip.

No, he said—he had called Hilbert from Oslo, but he had not gone to Göttingen.

When he arrived at his hotel in Carlsbad, he was informed that a man had telephoned, asked when Courant was arriving, and then hung up without leaving his name. Courant was certain that the caller had been Friedrichs, and the next morning Friedrichs appeared and announced— Courant told me—"I will even go as a dishwasher."

"I don't remember saying that," Friedrichs said with a smile, "but, well, Courant likes to tell it that way."

Courant returned to America in September 1936 full of plans for the future. In Carlsbad he had also met David Sarnoff, the president of RCA— "one of the most interesting men I have ever become acquainted with," he reported to Franck—and the acquaintance must be pursued, both for himself and for Franck. People must be contacted about an invitation for Artin. Preparations must be made for Friedrichs's arrival, some sort of position found. He had decided that it would be good for Flanders, who seemed to have lost his scientific and personal impetus, to spend part of his upcoming sabbatical year with Harald Bohr in Copenhagen—and that too must be arranged. Anna Stenzel would need support. The new rooms of the Graduate Center must be furnished.

Back at NYU, while he was still installed in his old office at the Heights, a little Jewish girl presented herself with a letter of introduction from Flexner. Courant looked doubtfully at her and asked if she could type and take shorthand.

"I *could* type, but I never learned shorthand very well," Bella Manel, now Kotkin, confessed to me, "but since Courant spoke very haltingly in English and since I was not afraid to ask him to repeat if I missed something, I was able to manage. After I gave my first report in the seminar, his attitude toward me changed completely. I was accepted then and asked to lunch with him and other students."

She still remembers very well Courant's first days in the little rooms upstairs at 20 Washington Square North. She acted as his secretary and receptionist, answering the telephone, giving instructions to students, supervising the small library—which consisted mainly of Courant's books and journals and his personal collection of reprints. People from all walks of life came to the office, she told me—famous scientists "and even royalty." It was all very exciting to a young student, but it was a relief when in the spring Courant got her a Blumenthal Fellowship, and she could begin to concentrate on her dissertation.

Sometimes on weekends she and other students, including Max Shiffman, to whom she was later married, were invited out to New Rochelle. There they were put to work doing chores in the yard.

"It is true that Courant utilized people to do things for him," she conceded, "but never without giving something in return. We were welcomed into the warmth of his family by him and Nina. That was wonderful, and it was worth a lot to us."

With his appointment permanent, Courant began to try even more intensively to bring his experience in Göttingen to bear upon the situation at NYU. During the fall semester 1936–37 he offered a general course for mathematics teachers as well as for graduate mathematics students, which he called "Elementary Mathematics From a Higher Viewpoint." In his opinion, Felix Klein's lectures under this title had contributed greatly to the improvement of secondary-school mathematics teaching in Germany and had also given mathematics students there a broader understanding of the connections between higher mathematics and elementary problems and between the different branches of their science. Something similar might have a similar effect in America.

That same fall G. H. Hardy, who was then a visitor at Princeton, came to NYU and gave some talks. He was much impressed by the responsiveness of his audience and later wrote Courant to that effect. Tremendously pleased and encouraged, Courant sent copies of Hardy's letter to a number of people he was trying to interest in the Graduate Center.

He continued to be fascinated by Plateau's problem and began to perform soap-film experiments which were more sophisticated than those which had been performed by Plateau.

For such experiments a thin wire frame, shaped to a given closed contour, is dipped into a viscous liquid similar to soapsuds and carefully withdrawn. If one ignores gravity and other forces which interfere with the tendency of the film to assume a stable equilibrium by attaining the smallest possible area, the film that then spans the frame is the physical representation of the minimal surface for that particular contour.

Although the physical existence of a solution to a physical problem does not establish the existence of a solution to the corresponding mathematical problem—as Weierstrass had pointed out in connection with Dirichlet's principle—the recent work of Douglas and Courant had established mathematically that such a surface does exist for every closed contour in space. Many other interesting mathematical questions about minimal surfaces were, however, still unanswered—among them, questions concerning the uniqueness and continuity of solutions.

Courant began to explore some of these questions by means of the soap-film experiments. Most of these are difficult to visualize without demonstration, but the description of a few can give some idea of the way in which the physical solution can suggest and sometimes answer mathematical questions. One wonders, for instance, if for any given closed contour there is only one possible minimal surface and discovers that a frame shaped like a headset (circular earphones joined by a curved band to go over the head) permits not one but three minimal surfaces. Another question concerns the effect of deformation on a contour. For such experiments little handles have to be attached to the frames. Then, after one has dipped a circular frame into the viscous liquid and drawn out a simple two-sided disk, he can gently give the frame a slight twist. Suddenly the surface of the film becomes a one-sided Moebius strip. Further deformation returns it equally suddenly to a two-sided strip. To explore other questions concerning the effect of free boundaries, Courant utilized frames which were flexible in part.

He found "the scope and informative value of soap-film experiments with minimal surfaces [much] wider than the original demonstrations by Plateau." For the next four or five years he was to play with them—in the seminar, in lecture halls, in his office, at home—with the absorption of a child.

The experimentation with soap films was only one example of a quality of playfulness in Courant—what Friedrichs described to me as "the ability to be fascinated." Although he could never understand his friend Hardy's passionate interest in cricket, he himself bought one toy after another—cameras, cars, phonographs, radios. Later he had the most sophisticated "hi fi" and tape recording systems.

Automobiles, beginning with the "thoroughbred" Röhr in Göttingen, were a particular delight.

"Courant was always having amusing, unique accidents," Friedrichs recalled. "Once, I remember, on a road in the mountains where the traffic was permitted to go up in the morning and down in the afternoon, he stalled his car for most of the day, half way up."

Hans Lewy conjectured that Courant played so intensely with his toys because he had never had any in his childhood.

During 1936–37, while Courant was giving lecture-demonstration on soap films at various eastern colleges, news of Göttingen continued to come from time to time. Otto Toeplitz—who with other Jewish professors remaining in Germany had recently been arbitrarily retired—made what he described to Courant as "a sentimental journey" to visit Hilbert at the New Year. Hilbert's memory, which had been greatly affected by his illness, seemed to Toeplitz to have improved.

"He definitely became the most animated when I told him about your activity. He wanted to know everything about it, and I could not tell him enough. . . . The liveliness of his interest in this part of my account which concerned you was the most lovely proof of how very much he was attached to you. When I just in passing mentioned that you had now done some beautiful things in the calculus of variations, he parried immediately, 'With that I have not concerned myself at all.' "

In closing, Toeplitz reminded Courant not to forget Hilbert's seventy-fifth birthday on January 23, and added: "About what has become of his enterprise in Göttingen, he is perfectly clear."

But for the most part, by 1937, Göttingen was in New Rochelle. "It seems to us that it is you who are in exile," Nina had long ago written to her mother, "and we who are at home." Time after time the house was filled with old friends, visiting colleagues, former students—such occasions always described by Nina as *sehr göttingisch.*

Anna Stenzel had been added to the household now in place of Hilde Pick, who had found "a real job." Courant put her to work typing the third chapter of Courant-Hilbert II. Suddenly, after thirteen years, the book was beginning to move. By December half of the manuscript had gone to Springer. In the spring Stenzel's daughter was sufficiently confident of her "American English" to take over from Bella Manel at the Graduate Center.

With both his writing and his mathematical work going well, Courant could still not escape the feeling that he should try to do something more for the situation of mathematics at NYU. Unlike many private universities, it had no significant endowment. Almost all expenses were met by student fees. Clearly, any new development could come about only with financial support from the outside.

There was already in existence a model for Courant of what could be done. The interest of public-spirited New Yorkers in art had been utilized by Walter S. Cook to provide NYU's Institute of Fine Arts with the funds to hire a stellar group of art historians who had been forced to leave Germany. Courant was optimistically certain that if New Yorkers would support art, they would also support mathematics which, in addition to being an art, was of uncontested practical value. Whenever an opportunity offered, he tried to establish contacts with wealthy and influential people, obtaining introductions, parlaying one acquaintance into another, presenting himself to virtual strangers by some tenuous connection. A sample of his approach in these first years is contained in a series of letters to a Mr. Henry Gaisman, to whom he had been "encouraged" to write by Percy Straus, the president of Macy's.

Mr. Gaisman, according to Courant's description of him, was over sixty years old, "a little bit bashful, very wealthy and sometimes generous." Starting out as a newsboy, he had become a successful inventor and businessman.

"I need not say, indeed, how delighted I should be if you would permit me to discuss matters with you," Courant wrote to Gaisman. NYU had a number of gifted students—"many Jewish, earning their living by hard work, some of them original personalities." One of his students was an ingenious inventor. The whole situation was such that "with help from the outside on a comparatively modest scale an enormous progress of high public usefulness could be achieved."

When after two weeks there was still no reply from Gaisman, Courant wrote again:

"I suppose that a man like you must be subject to every kind of pressure to contribute to various things of different merits. Therefore I wish to assure you that although I hope to enlist your interest, I certainly shall not ask you for any active help which you should not deem appropriate on your spontaneous judgment. Asking you for a chance to discuss the matter personally might already appear as some kind of intrusion; but the cause seems to me sufficiently worthy from many points of view, not the least of which is the Jewish angle. It is this consideration which makes me overcome the natural hesitation I feel about asking you again for the favor of a personal discussion."

For more than half a year Courant persisted in trying to interest Gaisman. During this time the only response he elicited was a note from Gaisman's secretary to the effect that his employer gone away for the winter.

For a while Courant also had high hopes of obtaining a grant from the Carnegie Foundation after his talk the previous year with the friendly Keppel, but this too failed to materialize. Somewhat hesitantly, he turned finally to Warren Weaver.

"I really do not feel impelled by personal ambition to embark again on a work of organization as I did in Göttingen . . . ," he assured Weaver in December 1936. "But seeing the need and the chances for the proposed development and feeling my deep indebtedness . . . I simply have to try. . . ."

The ultimate goal at NYU should be "a strong and many-sided institution . . . which carries on research and educational work, not only in pure and abstract mathematics, but also emphasizes the connection between mathematics and other fields as physics, engineering, possibly biology and economics, and which cooperates in helping to develop better standards in high school instruction."

A "highly efficient" use could be made of any money, even a small amount. For the library $4000 would be sufficient to make it satisfactory, $6000 would make it really good. "But also less than $4000 would be a great help."

Weaver was, by virtue of his own background, very sympathetic to Courant's ideas. He had taken a degree in civil engineering and had then done his advanced work in mathematics. His most recent teaching position had been as head of the mathematics department at the University of Wisconsin. In 1932, when Max Mason, whom he admired more than any other scientist, had asked him to come to the Rockefeller Foundation, he had taken a cool look at his abilities and concluded that he lacked "that strange and wonderful creative spark that makes a good researcher. . . . There was a definite ceiling on my possibilities as a mathematics professor." He had accepted Mason's offer.

Weaver was convinced that Courant had a "sound and important" plan:

"There are few places in the country where applied mathematics is being emphasized in any adequate and competent way; and a fine development of this sort in the New York area would seem to be clearly indicated."

Unfortunately a development of the kind Courant had in mind was, as Courant himself knew, "quite outside of the program and possibilities of The Rockefeller Foundation." In spite of this fact, a small grant to NYU was subsequently made and regularly renewed for a number of years.

"We make such exceptions only in cases which are judged, on the basis of the best evidence we can gather, to be of very unusual merit and importance," Weaver was always to remind Courant. "You are therefore justified in viewing this assistance, even though of very modest amount, as a real evidence of our interest."

One of Courant's main concerns when he talked to Weaver was that without additional funds existing chances in personnel might slip away. Among these he included J. J. Stoker, the assistant professor of mechanics at Carnegie Tech.

Courant had finally met Stoker in December 1936 when that young man had come to New York and given a talk before Courant's seminar. The interest, aroused by Hopf, had been immediately confirmed. Stoker was the son of an immigrant Englishman who had worked his way up from ordinary miner to superintendent of all the coal mines of Bethlehem Steel and had later become a mining inspector for the state of Pennsylvania. Young Stoker had originally studied engineering as it related to coal mining and after graduation had worked for a year as a mining engineer before returning to Carnegie to teach mechanics. There he had shortly come to recognize a serious lack in his education. In four years of college he had learned nothing of mathematics except a little elementary calculus. He decided that he should go back to school and study mathematics as it related to mechanics and physics.

At that time there was no university in the United States which offered the kind of work he wanted, Stoker told me when I talked to him in his office at the Courant Institute, which—if it had existed then—would have suited his purposes exactly. A few individuals at different places combined mathematics with applications. One was T. L. Smith, with whom he had taken some courses at Pittsburgh. Smith had studied in Göttingen after getting his degree at Harvard, and Göttingen was where he advised Stoker to go for what he wanted.

"And I would have gone except that by then—it was summer 1932— it was obvious that things were not good in Germany."

As an alternative to Göttingen, Smith suggested the Eidgenössische Technische Hochschule; and in what was probably the worst year of the Depression, Stoker took his pregnant wife, Nancy, and his two-year-old daughter and set off for Zurich to study with Ernst Meissner, an applied mathematician who had been the first new friend Courant had made during his student days in Göttingen.

Stoker planned to work with Meissner in elasticity, a subject with which he was already somewhat familiar; but after attending lectures by the various professors in Zurich, he decided to work with Heinz Hopf instead. "I found Heinz Hopf's lectures and his whole way of doing mathematics—well, very very attractive." One of Hopf's courses that year was on differential geometry, which involves notions that also play a role in the theory of elasticity. "I had never studied it, but I found it so really beautiful—he did it so well—that I asked him to give me a subject in it for my dissertation."

Somehow—Stoker doesn't really know exactly how—after their meeting in December 1936 Courant managed to arrange that Stoker was hired as an assistant professor in the mathematics department of NYU's College of Engineering at University Heights. It was agreed that he would also give a regular course of lectures at the Graduate Center at Washington Square.

"And did you ever get to Göttingen?" I asked.

"No, I have never visited Germany at all. I have been in Europe a lot, but after the war I never had any desire to go to Germany. The Germans never harmed any of my relatives, I'm not Jewish, but I felt so angry with them, I felt why bother visiting places like that. On the other hand, I have a bad conscience about it, because the individual Germans are like anybody else and I have liked some of them very much. Anyway, I had no desire to go to Germany. And so I have never been to Göttingen."

At the same time Courant was arranging Stoker's employment at NYU, he was also trying to find a position for Friedrichs, who—in Germany—was setting up what he now thought of as his "escape."

Since Friedrichs was of an age for military service, it was necessary for him to get permission from the military to leave the country even for a short time. Fortunately he had a sister living in Paris, and he was able to apply legitimately to visit her during his spring vacation. To go to the United States, he also had to obtain a visitor's visa from the American consul. This required affidavits that he had residence and employment in Germany. He was sure that his landlady and the clerk at the Technische Hochschule where he taught (who wore a swastika prominently displayed in his lapel) guessed his plans, but they signed the required papers without a question. Everything costing money had to be done before leaving Germany, for he would be permitted to take no more than ten marks out of the country. As surreptitiously as possible, he purchased passage on a steamship that went from France to the United States.

He and Nellie Bruell, who was French by birth and thus had a French passport, had arranged that she would stay in Germany until he was safely out of the country. Then she would go to France and wait at her father's in Lyon until Friedrichs had obtained a position and could send for her.

In Paris, Friedrichs telegraphed Courant that he was on his way. He arrived in New York on March 4, 1937, quite penniless. Friends lent him money, and a few days after his arrival he sent a letter to Germany formally resigning his position as a professor. He took it with him to Princeton to mail so that it would not bear a New Rochelle postmark. Courant found him a room and paid him to help with the second volume of Courant-Hilbert, which was now finished except for the final chapter.

"So I was his assistant again. That was fine with me. Most immigrants to this country start at the bottom. I felt perfectly natural about it."

Courant wrote letters about Friedrichs's presence in the United States to everyone he knew who was interested in the development of applied mathematics. He emphasized the two years that Friedrichs had spent at the aerodynamics institute in Aachen and presented him as "a mathematician in the style of C. Runge." He was in fact so active on Friedrichs's behalf that even Hans Lewy began to be afraid that his efforts to place Friedrichs might jeopardize his own position at NYU.

"But you don't need to worry," Courant reassured Lewy two months after Friedrichs's arrival. "It is actually so that the Dean of the Engineering School, the Dean of the Graduate School, the Head of the Aerodynamics Institute, etc., are all of one mind that we should not let pass this splendid opportunity to cultivate applied mathematics."

In June 1937 Friedrichs received a temporary appointment at NYU as "professor of applied mathematics" in the graduate department. Half of his modest stipend of $3,000 came from the Rockefeller Foundation and the other half from private individuals whom Courant referred to as "friends" of the Graduate Center. As soon as Friedrichs had a position, he wired Nellie Bruell—now Mrs. Friedrichs—*to come!*

Thus it happened that by the end of his third academic year in the United States, Courant had at this side the two colleagues who would work with him until his retirement—and both of them had come to him by way of Göttingen.

TWENTY

THE YEAR 1937 saw the publication in Germany of Courant's last book to be written in German and the publication in the United States of his first book to be translated into English. Both books, as always with Courant, had stories behind their publication.

To Friedrichs's amazement, when he had arrived in March 1937, he had found the second volume of Courant-Hilbert already in proof—except for the last chapter.

"It was like a miracle!" he still marvels. "Of course it had been in his mind all the time; but the fact that I would now come and we would go over it together and fix it up and so on, and of course write the last chapter —that was what did it."

I had heard references to the seventh and final chapter in the Courant family—how little Lori had once stamped her foot and cried, "Never mention that seventh chapter to me again. It has ruined a year of my life!"— and so it was with interest that I asked Friedrichs what had happened.

It had been planned from the beginning, he explained to me, that the concluding chapter would present a general existence theory for the solutions of elliptic partial differential equations. When the chapter had first been discussed, Friedrichs had been Courant's assistant. Their thinking on the subject had been very close. In the years since then, Friedrichs had moved toward an increasingly abstract approach to the ideas which were to be presented in the chapter. In 1930, while reworking a paper which he had written earlier in the general area of Hilbert's spectral theory, he had come across von Neumann's basic paper on Hilbert space, in which Hilbert's spectral theory becomes a special case in a general theory. That paper had been a revelation to him, and he had immediately translated his paper into von Neumann's Hilbert-space language. When he had returned to Germany after his visit to Courant in 1935, he had seen still another approach to the whole complex of problems, also involving Hilbert-space methods, but going much farther than he had gone in his earlier work.

"So when I came back in 1937 and started to work with Courant on the last chapter of Courant-Hilbert II, I had a conflict. I didn't think it should be written the way Courant saw it—in the style of 1925 or 1926, say. On the other hand, I knew that to write it the way I saw it would be unnatural for Courant and it wouldn't then be Courant's. So we discussed it. Courant insisted he had already done it my way. He hadn't really. He had some basic notions of Hilbert-space theory, and he recognized the importance of von Neumann's work. But it was not natural for him."

The two men struggled into the summer of 1937 with the chapter, working on the side porch of the house in New Rochelle. Courant had a great deal of trouble settling down; but, once settled, he could go on long after Friedrichs was exhausted.

"Richard sighs and groans and is very irritable," Nina reported. "From week to week he says, this week it must now be finished. But then it is not."

In July the landlord proposed to sell the house. Horrified at the thought of having to gather up manuscript and galley proofs and notes, Courant promptly bought it.

Finally, in the middle of the summer, Courant announced to the family that the seventh chapter was at last finished. To celebrate, he took them all to Radio City Music Hall. They had a wonderful time, but then in the middle of the night they heard him prowling around again, muttering to himself. It was all wrong, it was hopeless, it wouldn't do.

In the end the published chapter was a curious compromise.

"People always shook their heads about that chapter," Friedrichs said. "It was clear it wasn't just Courant's, but it wasn't just mine either."

Oddly enough, in the published book there is no mention of Friedrichs in spite of his dozen years of association with the project.

"That was on purpose," he explained. "The book was being published in Germany, and we thought it would be better not to associate me with Courant. You see, I still had family in Germany. It probably wouldn't have made any difference, but at that time we all had a tendency to overreact."

In spite of his delicacy in not mentioning Friedrichs's help, Courant audaciously proposed to dedicate the second volume of Courant-Hilbert to Harald Bohr, who had been the anathema of pro-Nazi mathematicians since his response to Bieberbach's speech on "Personality Structure and Mathematical Creation." At the last moment he gave up the dedication to Bohr because it might make difficulty for Springer.

The other book by Courant which appeared in 1937 was the American edition of the English translation of his calculus—the first of his books to be published in direct contact with a publisher other than Ferdinand Springer.

Springer had given the English-language world-rights to Courant when he left Göttingen for Cambridge in 1933. During that unhappy year Courant had given them to Blackie, an English publishing firm in which a Cambridge professor who had been friendly to him had had an interest. Blackie had no American co-publisher, and Courant had taken on the job of approaching various American publishers about issuing the book in the United States under an agreement with the English firm. He had considered making McShane, by then a quite well-known young American mathematician, a co-author. He had then abandoned the idea.

I asked McShane if he had ever received any royalties from the translated version of the calculus. He said no. The agreement between him and Courant had been that he would be paid for the translation and would get credit for it. "I never felt that Courant owed me anything," he said. "But he did support me considerably in my early career as a mathematician."

In 1936, while Courant was still engaged in dickering with American publishers about the calculus, he met Erich Proskauer, the editorial adviser of a German publishing house. Proskauer was in America to investigate the possibility of founding a European-style scientific publishing company in that country. The result was the organization in 1937 of the Nordeman Company (which in 1940 became Interscience Publishers). Although Courant had just negotiated successfully with a well-known American publisher to issue his calculus, he decided to give the book to Nordeman instead.

Proskauer, with whom I talked one morning in 1975 in his apartment facing the southern end of Central Park, is still amazed at "the mixture of naiveté and shrewdness" with which Courant gave his book to the new company. One attraction, he now thinks, may have been European attitude toward publishing which Nordeman represented. "In Europe the author was king while in the United States the reader was king."

The European viewpoint may have been especially attractive to Courant at that time; for he was having an unpleasant experience with the publishing of his first long paper in an American journal. Proofs had been returned to him by Solomon Lefschetz, one of the editors of the *Annals of Mathematics*, who had let up a howl at the "extraordinary" number of corrections. In addition Courant had asked for page proofs. Page proofs! "Owing to the delay that all this is causing us, we may be obliged to postpone your paper. . . ."

Proskauer smiled wryly when I recounted Lefschetz's reaction, remembering a time much later when the first five chapters of the English translation of Courant-Hilbert, already set up in type, had to be killed.

The agreement with Nordeman represented an important change in Courant's relations with his publisher. After the experience with Blackie, he was "very deliberate," according to Proskauer, and "very loathe to give up any of his copyrights." In addition to their agreement regarding Blackie's rights, he and Proskauer signed another agreement according to which author and publisher shared equally in profits. It was under this agreement that the first volume of Courant's calculus appeared in the United States in 1937.

There were by this time quite a few outstanding students at the Graduate Center. The little library had been augmented by the addition of a number of volumes belonging to Emmy Noether. Flanders, in Copenhagen for the year, was missed; but Friedrichs and Stoker furnished the companionship and support which Courant felt he required.

The two younger men—both then in their thirties—were as different from Courant as it is possible to imagine, and as different from each other. Stoker was straightforward, outspoken, "a moralist" in the words of later students and colleagues. He shared Courant's educational and mathematical philosophy and his sense of mission, but from the beginning he struggled against Courant's oblique way of operating. Friedrichs, who came from a lawyer's family, was judicious and, as Courant often complained, "legalistic." He had grown away from Courant mathematically, but he had a sympathetic understanding of "Courant's way" and was able to accept it. Still, even Friedrichs felt a need to protect himself if he was to survive. After he was married and living near Courant in New Rochelle, he established definite rules about times when he was available.

"My way of handling the problem was so alien to Courant that he was constantly annoyed by it," Friedrichs told me, "but he somehow or other knew it was necessary for me."

It was typical of the flexible organic development which Courant envisioned for the Graduate Center that in those early years Stoker, who was an applied mathematician, was "professor of mathematics" while

Friedrichs, who in spite of his two years with von Kármán was basically a pure mathematician, was "professor of applied mathematics." Both men subscribe to Courant's thesis that there is no boundary between "pure" and "applied," but they frequently use one word or the other quite specifically in conversation. I asked Friedrichs once how they made the distinction.

He replied that the distinction is made in different ways at different places by different people. He cited as an example two courses which he and Courant taught during those first years.

In Courant's lectures on mathematical physics, he discussed mathematical problems which had arisen originally from problems in physics. But he handled these problems in a purely mathematical way. He was concerned with general theories, with existence and uniqueness of solutions rather than with special methods of determining solutions concretely. His treatment was "pure," although he also referred the mathematics to the applications throughout the course.

"Now a very pure mathematician will call mathematical physics applied mathematics regardless of how it is taught, so to him Courant's course would be applied mathematics. We who work in the field call it pure mathematics."

In Friedrichs's lectures on fluid dynamics, on the other hand, although he formulated the problems mathematically, he discussed a number of special solutions—"technical methods for getting the answer"—and rarely tried to prove anything rigorously but simply referred the applications to the mathematical theory from time to time.

"I considered that what I was doing on fluid dynamics was applied mathematics. Of course somebody from the engineering department would probably have considered my treatment pure mathematics. So you see what I mean—it depends on who is making the distinction."

That same fall of 1937 when Friedrichs and Stoker joined Courant at NYU, Artin and his family finally arrived in the United States.

The mystery of Artin's long silence in regard to several American invitations had been cleared up when Flanders, in Copenhagen, had received news of him from a Danish mathematician who had been in Hamburg. The German ministry of education had turned down Artin's application for a leave of absence during the summer of 1937 on the grounds that he was indispensable. Other German mathematicians had urged him to remain in Germany. Among these was Hasse, who had recently applied for membership in the National Socialist Party.

"My endeavor at that time was to keep up Göttingen's mathematical glory," Hasse explained to me. "For this I needed the consensus of party functionaries at the university whenever I wanted to get some distinguished mathematician to fill a vacancy in Göttingen. Among these functionaries I had one close friend and one who was leaning towards helping me. They asked me to join the party so that they could help me better. It is true that I gave in and applied for membership. But on my application I put that there was a Jewish branch in my father's family. I was almost sure that this would lead to my application being declined. And so it was. The answer, which I received only after the outbreak of the war, was that the application was not going to be acted upon until the war was over. In the meantime, however, I had been able to help several mathematicians who were having political difficulties at other universities, by offering them positions in Göttingen."

On a visit to Hamburg in 1937, Hasse suggested to Artin that it might be possible for his children, who were only a quarter Jewish, to be declared Aryan.

The Artins lived at the end of a cul-de-sac; and whenever Natascha Artin, who was half Jewish, heard the sound of an automobile at night, she was sure it carried the Gestapo. But Artin, who still held his professorship in Hamburg, could not bring himself to accept the only permanent American position which had been offered to him—a professorship at Notre Dame. He had been born a Catholic, and he feared that the Catholic fathers in South Bend would find this out and try to bring his children into the church. In this situation—only a few months after his request for a temporary leave had been denied on the grounds that he was indispensable—he was removed from his professorship.

As soon as Courant had heard these details from Flanders, he had immediately suggested that Flanders write to Chancellor Chase, explaining the situation and the desirability of a "non-sectarian" position for Artin. Chase, who had already provided a place for Friedrichs that fall, had not felt he could take on another refugee. Artin had had to accept the Notre Dame invitation.

When Artin and his family arrived in New York on October 1, 1937, Courant and Hermann Weyl were at the pier to meet them.

"I will never forget our welcome by Courant and Weyl," Natascha Artin Brunswick told me with a smile at the memory. "You cannot imagine a greater contrast—Courant very little and not the sort of man you would look at twice—I think his face became much more interesting with age, and suffering too, I suppose—and Weyl, who was most impressive— a very, very interesting-looking man."

The Artins accompanied Courant out to New Rochelle, and a few days later Courant decided to give a party to introduce his friends to some of the big people in the administration at NYU. On the day of the party he and the Artins returned from the city about four o'clock. They found no one at home. Absolutely nothing had been done to prepare for a party. The house had not been cleaned or even straightened, and there was practically no food in the place. Courant was beside himself. Then finally Nina arrived, completely unperturbed, and said when he began to sputter. "But, Richard, you know this is the day I have my rehearsal."

Everybody, including the Artins, pitched in. By the time the guests arrived, the house had been vacuumed and food purchased and prepared. The party went off very well.

A few days later the Artins left the Courants' and went to Princeton, where there was to be another party in their honor at Weyl's home. His was "a very polished household"; and his party, an elegant affair with literary luminaries and famous philosophers as well as scientists among the guests. The country lad of Courant's student days had become a sophisticated man of the cultural world.

It was a great satisfaction to Courant to have Artin in the United States, and whenever that mathematician was in the east he regularly lectured at the NYU Graduate Center.

"The charming Artin family has been here," Nina wrote at Easter 1938. "He is teaching at a university in the Middle West . . . but Richard intends to get him to Princeton or someplace like that as soon as possible."

Courant spent the summer of 1938 writing what he described to Max Born as "a little book for teachers, not really popular, on higher mathematics from an elementary viewpoint. Very different from Klein's book and, as I think, quite necessary."

By September 1938, the beginning of Courant's fifth year at NYU, the activities of the Graduate Center of Mathematics had burgeoned into such an operation that more space was required than the tiny rooms on Washington Square North. Then, just as the semester began, Courant received a notice that even these rooms were to be taken from him—mathematics did not really require physical space beyond the lecture hall! Courant was very upset, but he still felt helpless in relation to the giant university. Thanks, however, to the intervention of a friendly young law professor, who went before the administration to plead his case, he was given some other rooms—a "suite" in a girls' dormitory next to the Judson Church on the south side of Washington Square—on the condition that the mathematics faculty and students would pass with appropriate decorum through the lobby of the dormitory on the way to their quarters.

Bedrooms were promptly transformed into offices, the living room into a library and lounge for the meetings of the mathematics club, a bathroom became a kitchen as well. In the long hall in front of the elevator, a ping pong table was installed.

A student who first saw Courant in these quarters was Charles De Prima, now a professor at the California Institute of Technology. In the fall of 1938 De Prima was twenty years old, the son of Italian immigrants, a boy who had grown up across the Hudson in New Jersey. After some college and work experience, he had come to NYU vaguely planning to go into journalism. He took a few courses in mathematics and physics, which had always fascinated him. At the suggestion of one of his teachers he went to a seminar talk at the Graduate Center, where Courant introduced the speaker. De Prima had heard Courant's name and had studied calculus from his book, but he had never seen him before. Yet somehow Courant seemed very familiar.

"*How* did I know him? Well, in the middle of the lecture it occurred to me that he was typical of the drygoods salesman who used to go from house to house at that time in the mixed Italian-Jewish neighborhood in which I had grown up. I had seen that man over and over again all my life. *Courant was that man.*"

After the lecture, when De Prima was introduced by his teacher, Courant mumbled something to the effect that the young man should come and see him sometime. Taking the remark as an instruction, De Prima turned up a few days later at Courant's office in the Judson Dormitory.

"He didn't remember who I was, but we started to talk. After a while he said something like, 'Well, fine. Do you know New Rochelle?' So vague arrangements were made that I should come out to his house there in a couple of weeks, and I did that too."

In New Rochelle, while Courant picked up a few sticks and talked, De Prima chopped wood, mowed the lawn, cleaned up the front and back yards, and found himself "simply entranced" by the way Courant talked about mathematics and Göttingen.

"It was so darn human. I had always thought of mathematics as something where you sit down at a desk and apply yourself. But now I found out you could do it while cutting grass."

De Prima, who left NYU relatively early in his career but has always remained in touch, has a particular perspective on Courant. In the course of our interview in the faculty club at Caltech, he talked to me about the influence of the second volume of Courant-Hilbert, which had just appeared when he was a student. Now, more than thirty-five years later, he feels that something happened with the second volume which was similar to that which happened with the first volume: "Only this time it happened for mathematicians rather than for physicists."

For some time, there had been a steadily growing and worldwide development in mathematics which emphasized abstraction, generality, and axiomatization. All his professional life Courant was to oppose this development, not because he opposed the aspects of mathematics which it represented, but because he felt that one side of the subject was being increasingly overemphasized at the expense of the other, equally essential side.

"Living mathematics rests on the fluctuation between the antithetical powers of intuition and logic, the individuality of 'grounded' problems and the generality of far-reaching abstractions," he said on the 100th anniversary of Hilbert's birth. "We ourselves must prevent the development being forced to only one pole of the life-giving antithesis."

In the subject of Courant's special interest, the trend to which he objected is seen in functional analysis, a development which received one of its early impulses from some of Hilbert's work during Courant's student days in Göttingen. Courant never liked even the name functional analysis. "What does it mean?" he would demand. "Analysis is analysis. You always deal with functions in analysis. What do you mean *functional analysis?*" And he would announce: "I do not specialize in generalities."

"I think he thought that freedom was being lost by the axiomatization of mathematics," De Prima said. "I think he was wrong in certain respects. I think that where initially you lost a freedom you gained others that you hadn't had before. But—anyhow—some of this had repercussions in the second volume of Courant-Hilbert." He looked cautiously at my tape recorder. "Maybe we ought to turn that thing off, because I am going to try to put into words something that I have never voiced before."

But he went on talking.

"You see, Courant-Hilbert II was one of the very few 'almost coherent' discussions on partial differential equations that existed in the late thirties and the forties. There were some others, but I think Courant was really the one who began to see the beginnings of a theory in this subject. It is so broad and so very loosely held together, yet certain things were standing out. Also there in Courant-Hilbert II, Friedrichs had written a chapter where the beginnings of the functional analytic approach had begun to show up."

"The notorious seventh chapter?"

"Yes, that's it. But it was done in a way that was, from the point of view of a functional analyst of ten years later, very clumsy. But what happened was that in the late forties people who were not at all trained in this Courantian way of thinking about partial differential equations suddenly saw the light through all of this and saw, my gosh, we have all the tools we need to attack these problems and give this thing some coherence. There were several important things that occurred during this period, and also some things later on at the Courant Institute. But by then it had taken a jump beyond Courant and maybe beyond Friedrichs too."

De Prima's comments have the advantage of more than thirty-five years of hindsight. At the time of publication, when Hermann Weyl was reviewing the second volume of Courant-Hilbert, he found it "comforting":

"When one has lost himself in the flower gardens of abstract algebra or topology, as many of us do nowadays, one becomes aware here once more, perhaps with some surprise, of how mighty and fruitbearing an orchard is classical analysis."

Shortly after the second volume of Courant-Hilbert was published in Germany, Courant began to talk to American publishers about the little book for teachers based on his lectures on elementary mathematics. By now he was considering *What Is Mathematics?* in place of Felix Klein's title, *Elementary Mathematics From a Higher Viewpoint*, and had broadened his idea of the audience for whom he intended the book to "the educated layman."

While Courant was thus occupied with publishing plans in the United States, scientific publishing in Germany was deteriorating very rapidly. In October 1938 Ferdinand Springer wrote to Neugebauer in Copenhagen, where Neugebauer was still editing several Springer journals, and instructed him to remove the name of Tullio Levi-Civita from the masthead of the *Zentralblatt* on the grounds that, as a result of Italian racial legislation earlier that year, Levi-Civita was no longer a university professor. Neugebauer promptly resigned his editorship of all Springer journals.

There were immediately many questions in America. Should the American editors—Courant, Tamarkin, and Veblen—also resign? Or should they, merely by threat of resignation, try to support Springer against orders they knew came from the German government? What of Neugebauer's professional future and what of the future of the *Zentralblatt*, which was the most important international review journal? Could the plan—in mind for sometime—of publishing a similar journal in America now be put into effect?

"I have the idea that pretty quickly you should come over here for a consultation and a few lectures," Courant hastened to advise Neugebauer. "That certainly is easily arranged—I have already some available cash on hand, which together with a few other invitations ought to finance the trip. . . . At the end of December there will be a big meeting of the American Mathematical Society, and it could very easily turn out that your presence there would result in a speedy solution of the problem."

After much correspondence the three American editors sent a joint letter of resignation to Springer. It was mailed rather than cabled because, in Veblen's opinion, a cable "would be a slight overemphasis."

The time had come, Courant now decided, for him to resign as editor-in-chief of the Yellow Series. During the past year, in agreement with Neugebauer, Bohr and Hardy, he had done everything possible to make easier what he recognized as Springer's "delicate position"; but recently contracts had been cancelled and plans dropped for books by a number of American authors. The end of such accommodation seemed to have been reached. Courant communicated his intention to resign to his fellow American editors, George Birkhoff and Marston Morse; but he did not ask them to join him.

Courant's letter of resignation to Springer was dated almost exactly twenty years after the two men had sat down in Berlin, in the midst of the post-war revolution, and had signed the contract for the Yellow Series. The books could now be found in every corner of the mathematical world.

(Two years later Max Dehn, fleeing the Nazis across Russia, went to inspect the library in Vladivostock during a train stop. In the entire library there was but one shelf of mathematical books—the Yellow Series.)

"I need not say to you how deeply this development, which lies outside your and my sphere of influence, disturbs me," Courant wrote to Ferdinand Springer on November 20, 1938. "But I would like to emphasize that my personal loyalty and friendship for you and my readiness to be at your disposal with my advice remain unchanged. . . . No matter how things may shape up, I will always look back with pleasure on our long collaboration."

TWENTY-ONE

THE ACADEMIC YEAR 1938-39 began with the Munich agreement. It was Courant's fifth year at NYU. He carried on his teaching duties, lecturing and conducting his seminar. He continued to concentrate on Plateau's problem and other minimal-surface problems. He worked with various students on *What Is Mathematics?*, paying them out of a $1500 grant he had obtained from the Rockefeller Foundation for that purpose. He also approached the Philosophical Society about a grant to write a book on Dirichlet's principle. He gave demonstrations of soap-film experiments and a talk on the radio on "Infinity." He participated in the early stages of the negotiations which resulted in Neugebauer's coming to America as editor of the new abstracting journal sponsored by the American Mathematical Society—the *Mathematical Reviews*. He continued to seek financial support for the library and for fellowships at the Graduate Center, to arrange to bring important mathematicians there to lecture, to try to broaden the base of interest in New York City as a whole.

It was at this time that Morris Kline, who had spent three years at Princeton as the assistant of the topologist J. W. Alexander, returned to NYU as an instructor in the undergraduate department at Washington Square.

"When I came back from the Institute for Advanced Study in 1938, Courant made it pretty clear that he couldn't afford to expand in many fields," Kline recalled, "and so I began to shift gradually to applied mathematics through my own studies, even during the period from 1938 to 1942."

I asked Kline why he thought Americans had always been so inclined toward pure mathematics.

"I think that when the Americans first became research-conscious they took up the newer fields which didn't require so much background; for example, they went into abstract algebra and topology. In the early 1900's those were new fields. One could understand what had been done without too much background. One could see any number of problems that were open. If they had chosen to go into analysis, they would have had to acquire much more extensive backgrounds before they could do original work of their own. I think that people in our country got started the easiest way at the time."

210

(Such a narrow purism was not characteristic of many of the outstanding native-born mathematicians. Birkhoff, speaking in 1938 of Benjamin Peirce—"a kind of father of pure mathematicians in our country," also spoke for himself: "In his deep appreciation of the elegant and the abstract we may recognize a continuing characteristic of American mathematics. In his concern with its many applications there resides a virtue which we are finding it more difficult to realize, because of the trend toward professional specialization. Without doubt, however, there is a spiritual necessity upon us today to regain a similar breadth of outlook.")

All of Courant's activity during the academic year 1938–39 took place against a background of involvement in what was happening to his friends still in Germany and to those who had already emigrated.

Except for his brother Fritz, he continued to brush off problems of relatives. But for Marianne Landau, who had emigrated after her husband's death early in 1938, he hastened to help, firing off a number of letters on her behalf. One of these went to Paul Muni, who had played her father, Paul Ehrlich, the discoverer of the cure for syphilis, in a motion picture. And for Otto Toeplitz, who was trying to obtain a visitor's visa to the United States in order to establish contact between Jewish charitable organizations in Germany and similar organizations in the United States, he offered in a letter to the American consul "to assume personal responsibility by posting bond if necessary, [and] to extend to Professor Toeplitz my personal hospitality."

Up until the beginning of November 1938, many Jewish professors, like Toeplitz, thought that although "emeritized" they could continue to live and work in Germany, since they still received their pensions. Then the government announced that on November 9-10 male Jews under sixty years of age would be arrested and imprisoned for a period. On that night, which became known as the *Kristallnacht*, the sound of the shattering windows of Jewish businesses and homes was heard all over the country. A few days later Courant received a telegram from Siegel announcing that Hellinger had been sent to a concentration camp. There was no news of what had happened to Toeplitz.

Two weeks after the *Kristallnacht*, the American Mathematical Society, which had been founded on November 24, 1888, as the New York Mathematical Society, concluded its first half century of existence. The anniversary had earlier been celebrated at a meeting of the society in

September. On that occasion Birkhoff, as the leading representative of native American mathematics, had summarized the history of the subject in the United States and Canada.

The extraordinary contrast between 1888 and 1938, he pointed out, was manifested by the increase in competent mathematicians in the country. Among these he directed the attention of his listeners to two special groups. One was those mathematicians who had shown "the rare quality of leadership" by their participation in the activities of the mathematical society; the other was made up "of mathematicians who have come here from Europe in the last twenty years, largely on account of various adverse conditions."

This influx had recently been especially large.

"And we have gained very much by it. Nearly all of the newcomers have been men of high ability, and some of them have been justly reckoned as among the greatest mathematicians of Europe."

Their coming had certainly worked to the advantage of the general mathematical situation in America. As a teacher of young American mathematics students, however, Birkhoff saw the continuing arrival of outstanding foreign mathematicians as a danger against which he felt he must warn his countrymen.

"With this eminent group among us, there inevitably arises a sense of increased duty toward our own promising younger American mathematicians. In fact most of the newcomers hold research positions, sometimes with modest stipend, but nevertheless with ample opportunity for their own investigations, and not burdened with the usual heavy round of teaching duties. In this way the number of similar positions available for young American mathematicians is certain to be lessened with the attendant probability that some of them will be forced to become 'hewers of wood and drawers of water.' I believe we have reached a point of saturation, where we must definitely avoid this danger."

The words were underlined in the minds of the refugee mathematicians who heard Birkhoff. Many of them were sure that "foreigners" meant "Jews." A number of Americans, including many who had no Jewish blood, resented the distinction between "American" mathematicians and those equally "American" but born elsewhere. Birkhoff himself was only two generations removed from Europe. But many quietly applauded.

In Birkhoff's defense it should be said that the situation of gifted young American Ph.D.'s in mathematics was desperate in 1938. Many were married and already had children. The lucky ones were National

Research Fellows, who received $2400 a year. Those who had been unfortunate enough to choose fellowships over jobs when jobs were still available now tried to live on stipends of $1000 a year at the Institute for Advanced Study. In some colleges there were a few "teaching assistantships" which paid $750 a year. The less gifted, of course, did not have jobs of any sort.

I asked Courant if he felt that Birkhoff's remarks about foreign mathematicians were a result of the fact that he was generally considered, as someone tactfully put it in a memoir after his death, "not pro-semitic."

"I don't think he was any more anti-semitic than good society in Cambridge, Massachusetts, used to be," Courant objected. "His attitude was very common in America at that time. I think Birkhoff was narrow and certainly he was wrong—but he was a very good mathematician."

A glimpse of the situation of Jewish mathematicians in Germany after November 10 is given in the communications of Hermann Weyl, chairman of the German Mathematicians Relief Fund, which he had organized in 1934 with Emmy Noether.

"According to an alarming telegram, Alfred Brauer (Berlin) seems to be in serious danger in connection with the recent pogrom," Weyl told the other members. "In an attempt to bring him over to this country at once, I pledged $300 from the Relief Fund and beg you to endorse my commitment."

This left a balance of $43.75 in the treasury.

In less than a month an assistantship for Brauer had been arranged at the Institute for Advanced Study.

"[And] we have invited Hellinger, Hamburger, and Arthur Rosenthal to come over without a stipend," Weyl now reported. "Affidavits have been prepared for these men and also for Dehn. Dehn obviously will try first to go to England or the Scandinavian countries. It is sure that Rosenthal and Hellinger are in concentration camps; probably also Hamburger. We are in touch with [Hellinger's sister], Siegel, and Mrs. Dehn. Nothing has been heard of, or done for, Toeplitz. Remak is in a concentration camp but we could see no way of helping him; however, I am told that the English will try to do something for him. . . ."

A month later Courant received a communication from Toeplitz, who did not make any mention of his experiences on November 10. He wrote merely that the trip to the United States would have to be given up. The American Consul kept making new stipulations "and does not have the courage to say no."

"We thought we were unlucky when Richard was one of the first professors to be dismissed," Nina said to me, "but we were the lucky ones."

After the *Kristallnacht* older mathematicians like Dehn had to be got out of Germany, too. The need to help the younger, unestablished men continued. Pleas for help were also now coming from places other than Germany, especially Italy. Even people who had already successfully emigrated to the United States were often without resources after their original temporary appointments had expired.

There were some particularly desperate cases. One of these was that of Eduard Helly, who had so disastrously interrupted Courant's report in the Hilbert-Minkowski seminar of 1907. Helly had become an actuary in his native Austria and as such had had more difficulty finding a job in America than even a young scholar. Someone sent him to Friedrichs, who had a small and temporary tutoring job available.

"I don't actually remember that it was Courant who sent Helly to me," Friedrichs says now, "but it seems most likely that he was the one. It would have been like Courant."

Then there were, as Courant wrote unhappily to Harald Bohr, "problems not connected with mathematics"—like the case of his brother and his family, who were still in Italy. He expressed the hope that a more liberal handling of immigration might permit scholars without promised positions to enter the United States, but he very much feared that for the masses of Jews in Europe there was no solution: "The future development of the present situation can hardly be imagined."

Christmas 1938 was not happy in New Rochelle. The children complained that Papa was dreadful about money.

"Richard is uneasy about money now for two reasons," Nina explained to her mother. "It seems advisable that his brother Fritz and his family leave Rome so—for a couple of years—they may be dependent on his support. . . . And then the responsibility for the people for whom Richard has given affidavits, and where he is in fact implored for still further guarantees to help them, burdens him also. He does not permit himself any concert or theatre—and would indeed hardly be in the frame of mind for it. He is often in a terrible mood."

The year 1939 began with a visit from Niels Bohr, and that brightened the gloom for Courant; for to him both Bohr brothers were kings.

In the course of Bohr's visit to the United States, he also took Courant with him to call upon his friend Lewis L. Strauss, a banker who was interested in scientific matters. Later, writing to Strauss, Bohr commented that for the most part the refugee problem was being treated by the Americans in a way helpful for the organization and development of scientific research in America: "In that connection it is of course very important that all sound undertakings of such kind find proper support, and I thought I might take the opportunity to express my great sympathy with the endeavors of Professor Courant to utilize his unique experience and the traditions of the former great German schools of mathematics in advancing the closest possible cooperation between pure and applied mathematics in America."

About the same time that Courant thus became acquainted with Strauss, Flexner asked him to summarize the changed situation of the mathematical-physical faculty of Göttingen since the Rockefeller Foundation's grant.

Siegel was currently in Göttingen with Hasse and Kaluza, Courant told Flexner. The places in physics had been filled by two "rather good" men, Becker and Joos, neither of whom, however, could be considered comparable to Franck or Born. Prandtl's aeronautics institute had been completely dissociated from the university and put under military control —the expansion was enormous. The Rockefeller Foundation's gifts to mineralogy and physical chemistry had been negated by subsequent Nazi dismissals.

"In judging the effect of change," Courant reminded Flexner, "one must not be deceived by the fact that still there are quite a few very good scholars in Göttingen. They are now isolated, mostly unhappy and dissatisfied individuals, and nothing seems to be left of the old spirit of closest cooperation and comradeship also with the younger members of the staff and with the older students."

By late 1938 a position at Brown, in addition to the editorship of the *Mathematical Reviews,* had been arranged for Otto Neugebauer. Courant could not have been happier as he went down to the pier in February 1939 to meet Neugebauer and transport him out to New Rochelle. Neugebauer was not only one of Courant's heroes, but also an intimate friend and longtime associate who would now be only four hours away by train.

"It is wonderful how more and more of Richard's friends are here in the vicinity," Nina wrote. "That pleases Puss in Boots and makes him purr."

At this same time, to make him even happier, Courant received the opportunity which he had so long desired—an invitation to spend a summer term as a visiting lecturer at the University of California in Berkeley. That spring of 1939, as Hitler occupied Czechoslovakia, Courant began to plan *un grand voyage d'instruction* for his family, crossing the United States by the southern route and returning by the northern.

At the end of April, Hitler denounced the German-Polish non-aggression pact and the Anglo-German naval treaty. In May he announced the understanding with Italy was now a "pact of steel."

Courant had tried everything he could think of to arrange a position for his brother Fritz in the United States. This had included approaching Donald Flanders's oldest brother, Ralph, who was at that time president of the Federal Reserve Bank of Boston. Fritz was, according to Courant's description, "a businessman of wide engineering experience and knowledge [who has] been extraordinarily successful." (Until recently he had been establishing and equipping railroads in Ethiopia and Libya.) But even with such recommendations and such contacts, Courant was not able to place his brother in the United States. In the spring of 1939 Fritz Courant emigrated with his family to Brazil.

In spite of the events of the spring, at the beginning of June when the Courants and their three older children set off for Berkeley, it did not seem to Courant that there would be war in Europe.

But the problems of the refugees pursued him even to what Nina described as "the magic land" of California. One of the most difficult cases was that of his gymnasium classmate Wolfgang Sternberg, who had emigrated but had been unable to obtain a position on the east coast. Courant felt that Sternberg was an extremely gifted man who had been prevented from realizing his potential by deep-rooted psychological problems. Since 1933, however, he had been constantly plagued by Sternberg's pleas for advice and assistance. He found repugnant any reminder of his youth in Breslau. Yet he remembered gratefully how the Sternberg family had invited him regularly to a *Freitisch*, the weekly meal an orthodox Jewish family shared with the poor. At the same time the memory of having been the recipient of a *Freitisch* made him very uncomfortable.

Because of Sternberg's personal problems, he felt that he could not recommend him for a teaching position; but before leaving Berkeley, he wrote kindly and at length suggesting that Sternberg come to California, "where the weather is better," and study applied statistics under Jerzy Neyman "and hope that something in connection with statistics and applications, not with teaching, might develop."

At other times he avoided Sternberg and did not answer his calls and letters.

There were some refugees who were difficult to help because of their critical attitude toward everything American and their feeling that a distinguished position was only their due.

"You do not seem to have any feeling what it is all about," Courant scolded one man. "The only safe method for you is: keep your mouth shut, never talk about yourself and your sad fate, never complain, never criticize anything, weather, climate, subways, women, administration in a country that has not asked you to come. . . . Forget your European-German superstition that scientific standing establishes a claim in society or from society. . . . This is a friendly warning which better comes now than later when it might be too late . . . you may accept or reject my statements. Discussion would be idle talk, for which I have neither inclination or time. . . . You are fairly well at the end of the rope, and it depends solely on your tact and wisdom whether the chance given to you now will result in making your talents available for useful service in society or whether you are heading for ultimate failure."

On August 5, 1939, writing a letter to her mother while sitting under a giant redwood tree in Muir Woods, Nina noted below the date: "25 years ago war broke out." A few weeks later, the Nazi-Soviet non-aggression pact was announced. While the Courant family was driving back to New Rochelle, they heard the news that the German army had marched into Poland. Two days later Great Britain and France declared war on Germany.

Ernest Courant remembers his father calling his tribe together and telling them that from that moment forward they were not to speak German in public.

It was shortly after this instruction that they were eating at a roadside restaurant. As Courant got up to go for a second cup of coffee, Nina called after him—in German—to bring her a cup too.

"I do not understand you," Courant replied softly but distinctly in English.

Nina repeated her request, still in German.

"I do not understand you," Courant hissed fiercely in English.

Nina asked a third time, still in German, but now loudly and disstinctly.

"I do not understand you!" Courant shouted. *"Du weisst ganz genau, was ich meine."*

At the time that war began in Europe, Courant had still not met the residence requirement for becoming an American citizen. But he felt like one. He was sure that the United States would be in the war eventually, and he wanted to do everything he personally could to help prepare the country. He was convinced that the kind of mathematics and the kind of mathematicians he was trying to develop at NYU would be needed.

Very shortly after they had come to NYU, Friedrichs and Stoker had begun a collaboration which represented for Courant exactly the kind of thing which he had hoped would occur in the kind of mathematical situation he was trying to develop. This work had originated when Stoker had discussed with Friedrichs a problem in Stoker's favorite subject—elasticity.

"Stoker knew the engineering significance of the problem, which involved the buckling of plates, much better than I," Friedrichs explained to me, "but we worked on it together. Worked very hard and got some results and developed a rather rounded theory for a special case. This was not exactly the realistic case. The realistic case was the buckling of a rectangular plate, we took a circular plate—it was mathematically simpler —but what we did clarified the whole situation. I had more mathematical background, Stoker had more engineering, and we worked very well together."

By the time classes began at NYU in late September 1939, the Germans had virtually subjugated Poland. One German triumph followed another. Courant worried about Siegel, still in Göttingen. Then in March a cable arrived from Harald Bohr saying that Siegel was in Copenhagen and eager to return to Princeton. The Institute for Advanced Study offered an emergency appointment for one year, and Siegel left Scandinavia three days before the Germans invaded.

Now it was the fate of the Bohrs that was of concern to Courant and their other friends. With a special grant, the Rockefeller Foundation made it possible to invite both brothers to America; but, although expressing their appreciation, they both refused to leave their institutions and their friends in Denmark.

Before the war began, Toeplitz had left Germany for Palestine. At the beginning of 1940, he died there. Hellinger had also managed to leave Germany, and he was in the United States by 1940.

On May 14 of that year, Courant wrote a long newsy letter to Born, who had remained in England. Up until the declaration of war, the two men had continued to carry on their correspondence in German; but now they wrote, still a little awkwardly, in English.

The eastern part of the United States was literally flooded with scholars from Europe, Courant told Born. Hellinger had at last "a very tiny and insecure position" at Northwestern.

"I sometimes think that Toeplitz's lot was perhaps not so tragic. He might easily have faced terrible disappointments all around, and none of us can really hope to live to see a somehow not completely crazy world."

Between the time that Courant wrote his letter and Born received it, Holland and Belgium surrendered to Germany. Nevertheless, on the day the letter arrived, Born had been prepared to answer with characteristic cheerfulness "in spite of the grave situation." Then, the next day, he heard that France had capitulated.

"Now there is no way of being cheerful, for even if America would come in immediately there would be not much hope. . . . I stand to the British as long as there is any possibility. But I must reckon with the other possibility that a particular fate is in store for us, Lublin or something equivalent. . . . But as nobody likes to go without any resistance I wonder whether you and our friends, Einstein, Franck, etc., could find a way out for [us]. . . . If you write that we shall live in a crazy world I am afraid that is an understatement—if we shall live at all."

During that first disastrous year of war in Europe, there was beginning to be more activity at the Judson Dormitory. The mathematicians were able to take over another suite from the girls. Since Courant's arrival more Ph.D.'s in mathematics had been awarded by him at New York University than in all its first century of existence. Although Friedrichs and Stoker and some other men from the Heights gave courses at the Graduate Center, it was Courant who seemed to be always around, talking to the students, asking them about what they were doing, telling them they must come out to New Rochelle sometime, becoming involved in their lives, realizing always that at times personal problems could be more important than mathematical problems.

He loved to tell stories of Göttingen. He told so many stories about Hilbert (who was still alive in Germany) that when someone facetiously referred in print to Richard Courant's *Complete and Unexpurgated Book of Hilbert Stories*, many people took the existence of such a book as a fact.

"I suppose every math department has its stories," De Prima reflected, "but Courant's were part of the attraction of NYU for me. I graduated in 1940 and with the war, the impending war, I didn't know what I should do. I applied to various places and got some acceptances, including even the offer of a fellowship to Princeton. Then I had this long session with Courant, and—that is how well he can sell things, you see."

"How did he 'sell' you?"

"Well, he is both obvious and subtle in these things. He kept pointing out what could be done at NYU. You see, the reason he was in New York— and he would hint at this on many occasions—why should he of all people stay in New York? He had no school, no staff besides Friedrichs and Stoker—and they were up at the Heights mainly. The thing was actually not a school at all. There were so many things that made it unreal, and yet Courant believed in it. He believed very strongly in it, because he felt that in the city—in the whole area—there was a tremendous amount of natural talent."

"A reservoir of talent?"

"Yes. That's the word he used. *A reservoir*. This tremendous reservoir of talent. He felt that there were many gifted people in the area—they might even be working, doing other things, but the reservoir was there, we had to tap it. Also Courant was keenly aware that war was coming, we were going to be involved. Even if we were not directly involved, we were going to have to do a great deal in this country scientifically. And then there were the stories about Göttingen. All these things just impressed me so. I felt so close to Courant. I had the feeling that I was part of the family. And so I stayed at NYU," he concluded, "and had the gall to turn Princeton down."

In the spring of 1940 Courant became an American citizen. He felt that he was now no longer a member of that class of "foreigners" who must always hold back their opinions. He began to speak out quite forthrightly about what he thought should be done in science to prepare the United States for the inevitable war with Germany.

TWENTY-TWO

URING 1940 and 1941 Courant was deeply concerned with two projects, both of which he conceived as contributions to higher education in America by one who was "profoundly grateful" for the opportunity offered him by that country. The first was a plan for a *national institute of basic and applied sciences*; the other was the book tentatively entitled *What Is Mathematics?*

Since his arrival in 1934 he had been pondering the scientific needs of his new country. He saw these as extending far beyond the capabilities of any single university. In the United States there should be a national scientific center—like the École Polytechnique in France—which would produce a responsible, well-trained elite of scientific manpower for the coming war and for the difficult years which would follow.

The first two drafts of a memorandum entitled "A National Institute for Advanced Instruction in Basic and Applied Sciences," are without dates, but the third is penciled "Winter 1940–41."

From the first paragraphs in all versions, the voice of Felix Klein rings out. Klein had intensely admired the educational philosophy and program of the École Polytechnique and had always regretted that the ideals it represented—pure and applied science in close connection, teaching and research combined, personal contact between faculty and students—"had never taken proper root in German soil." From the time he had arrived in Göttingen, he had endeavored to establish these admired ideals there.

In his memorandum Courant stressed the national situation out of which the École Polytechnique had developed. After the Revolution, France had been "economically disrupted, intellectually and morally uncertain, at war with the strongest powers of Europe, her educational facilities disorganized." Scientists "of vision and initiative" had conceived a plan for an institution of higher learning "at a level unheard of before," the students democratically but carefully selected, the best scientists in the country as teachers. The new facility had soon fulfilled "the highest expectations" of its founders; and within less than two years Army, Navy, Industry, and Government had begun to receive a supply of men whose education was superior to anything then available in the world—"men equipped for productive work and not dulled by a dead routine of training."

Courant's view of the national institute he proposed was a broad one— "something very universal," he told me, "not only mathematics but also physics, other sciences, and also history and philosophy." He did not see it as competing with the Institute for Advanced Study or with any other existing facilities. Rather it was "to supplement them, in close cooperation."

After working his ideas over several times in 1940—the changes were almost always in language rather than in content—he decided at the beginning of 1941, now that he was an American citizen, to circulate the proposal. It seemed to him that such an institute could be put into operation quite soon if it began simply with a program in mathematics and physics. At the end of the memorandum dated winter 1940–41, he optimistically set the opening for September 1941.

During the academic year 1940–41 Courant was also devoting time to the other project which he conceived as a patriotic service, the book *What Is Mathematics?* He had been working on it off and on for almost five years and had drawn a number of his students into the project. In addition to David Gilbarg, who wrote up the notes of the original lectures, seven young men, including Courant's son Ernest, are mentioned in the preface as helping "in the endless task of writing and rewriting the manuscript."

By spring 1939 Courant had apparently decided that the subject matter of the book was a little narrow, too restricted to his own interests. On a visit to Princeton, he had asked various people for recommendations for an instructorship which was open at the Heights. Marston Morse had suggested his assistant, Herbert Robbins, a young topologist from Harvard; and Courant had stopped by Robbins's office to meet him.

When I talked to Robbins in 1975 in his Riverside Drive apartment just outside the gates of Columbia University, he did not remember whether it was at this first meeting or one shortly afterwards that Courant brought up the subject of also working on *What Is Mathematics?* He told me that he came to NYU in the fall of 1939, taught elementary subjects at the Heights during the day and gave advanced lectures at the Square at night. Courant handed over to him what earlier assistants had already done with *What Is Mathematics?*, talked about the concept of the book, and asked him to go over the manuscript and improve and amplify it.

I asked Robbins how he and Courant had worked together.

"It's hard to say," Robbins answered. "He had a set of mimeographed notes which were a course he had given sometime previously, which were written up by someone who had taken the course, a student, and they formed about a quarter or a third of the material that finally ended up in the book. Some chapters were there in their final form, some weren't there at all.

So I would suggest, say for example, that we have a chapter on topology, and we would discuss what should be in it. On some things he had very definite ideas, and on some things I had very definite ideas. Over a period of two years I would just work away and show him the results; and he would comment and criticize and I would re-do them. . . . Sometimes he would think of clever ways of doing something, and sometimes I would. . . . I don't know that there was anything particularly striking about the method of collaboration. It was pretty close collaboration, although we never sat down to write together."

Robbins stopped to explain that he never took the course as Courant gave it, or any other course taught by Courant.

At first—Courant told me—he found young Robbins not much help— "in fact, he impeded me by not doing very much work and pulling a little bit in a different direction"; but then, after a frank discussion between them, Robbins settled down and was "quite useful."

Finally, as Robbins recalls, Courant told him that he was very pleased with the work to date. The $1500 from the Rockefeller Foundation, out of which he had been paying Robbins, who made $2500 a year as an instructor at NYU, had by now been used up. Courant suggested, according to Robbins, that the two of them make *What Is Mathematics?* "a joint authorship production," larger and more comprehensive than originally planned.

"Are you sure that that was what he said?" I asked, for I had heard many tales of people leaving Courant's office not at all sure from his mumblings what had been decided.

"I was never in any doubt," Robbins said emphatically. "My initial cooperation was largely because I wanted to make some money. The reason I wasn't so happy [at first] was that [the book] was taking a lot of my time, and for someone who's just got his Ph.D., as you know, his future reputation depends on research and not on popular exposition. So I was sort of hesitant to put in another year and a half or so of what I would regard as a distraction from what I was really interested in doing. . . . The original thing was merely . . . helping someone else with his book. I didn't expect to be a joint author. . . . But when he suggested making it a joint book, then I agreed. I too had gotten sort of engrossed in it by that time."

With the new plan, the book suddenly began to move. Proskauer advised Courant to seek a more general publisher than Interscience for the audience he wanted to reach; and at the beginning of 1941 Courant opened negotiations with McGraw-Hill, who had earlier expressed interest in *What Is Mathematics?* It occurred to him that the book might also serve as a wedge for interesting a big established American publishing house in some other ideas he had. Barely acknowledging a proffered contract, he began to sketch a plan for a series of advanced mathematical textbooks patterned on his Yellow Series.

While Courant was working with Robbins on the book, he was also concerned about the young man's future as a mathematician. He felt that although Robbins was very talented, he was not making the progress in his chosen field of topology which he should be making. (Robbins explained to me that he was not able to get his thesis into print until March 1941, because he was being thoroughly overworked at NYU and in New Rochelle.) The previous year an arrangement to have Will Feller, by then at Brown, give a course on probability and statistics at NYU had fallen through; and Courant had assigned the course to Robbins.

"I had just a few weeks' notice until it started," Robbins recalled, "and up until that time I had not the faintest acquaintance with or interest in either probability or statistics."

Courant was very impressed by the job Robbins had done with the course. He thought it would be desirable for the young man to study statistics and probability "at the source." The source, as far as Courant was concerned, was Jerzy Neyman, the famous Polish authority in the field, then recently come to Berkeley. In early spring 1941 Courant wrote Griffith Evans, the head of the mathematics department at the University of California, "with perhaps a strange kind of suggestion which I hope you will deem justified by the extraordinary general circumstances."

It was Courant's suggestion that Robbins, who was quite poor and the sole support of his mother and younger sister, be given some "modest" financial support by Evans so that he could work with Neyman during the summer.

"I am sure that this will be of great benefit for Robbins, and maybe for statistics."

Knowing that Robbins is now one of the outstanding men in statistics and probability, I asked him if he had in fact been given the opportunity to go to Berkeley in the summer of 1941 to work with Neyman.

"No," he said—he had not even known that Courant had made such a suggestion to Evans. "Had I done so, my life would have been quite different, I'm sure, because I didn't meet Neyman until much, much later."

During that spring of 1941, Courant was extremely busy. He was trying to finish the book, trying to interest people in his idea of a national scientific institute, trying also to set up at NYU a summer series of defense-oriented mathematics courses. Thus harried, when the people at McGraw-Hill expressed some slight skepticism about the commercial possibilities of *What Is Mathematics?*, although still insisting upon their willingness to to publish it, he was very upset. He was unwilling to make any concessions about the book's handling, and he was determined that it must be in print by the fall of 1941.

He had already got a taste of publishing in his profit-sharing arrangement with Interscience Publishers. Now he decided to become his own publisher. In the summer of 1941 he borrowed money from a wealthy friend and contracted with the Waverly Press to set the manuscript in type. He then signed a contract with the Oxford Press to handle the distribution.

During this same period Courant continued to press his idea of a national scientific institute. As the year 1941 progressed—the Balkans occupied, the British pushed back in North Africa, the Soviet Union invaded—it became in his view an *emergency* institute.

At first, rather than soliciting directly for funds, he tried to arouse interest in the project among well-known public figures who could provide him with access to additional non-academic support. One of those to whom he wrote was Justice Felix Frankfurter, whom he had recently met through Flexner. Another was Dorothy Thompson, the newspaper columnist, to whom he had been introduced by Donald Flanders's brother Ralph "about two years ago . . . in the Savoy-Plaza dining room." He asked her also for an early opportunity for a quiet personal discussion "to seek your advice in an educational matter that might become of non-negligible importance for the pressing problems of the day and for the future problems of reconstruction."

Justice Frankfurter responded regretfully. Dorothy Thompson's secretary gave him an appointment and then cancelled it.

What disappointed Courant most, however, was the lack of enthusiasm and cooperation on the part of the fellow scientists whom he had particularly counted on to support him, such as von Kármán, who—Courant told me—thought the institute idea was just "a cheap imitation of Felix Klein."

"Kármán knew Courant very well and he realized that Courant was not really an applied mathematician," Friedrichs explained. "Yes, mathematical physics—Courant-Hilbert, fine—but that is not applied mathematics. Kármám had come to the United States before Courant. He had the engineering background. He also had a very good understanding of *applying* mathematics—he wouldn't have said 'pure' mathematics but rather 'sophisticated' mathematics—to engineering. For all these reasons Kármán felt that *he* should be the one to develop applied mathematics in this country, not Courant; and he sometimes expressed himself in a not too friendly way about Courant. But, as you can see from Kármán's autobiography, he could not have done what Courant did. In fact, in later years he reluctantly agreed with me that this was so."

The most effective opposition to Courant's plan came in his opinion from R. G. D. Richardson, the dean of the graduate school at Brown University and one of the relatively few Americans active in placing refugee mathematicians. Since 1933 he had arranged positions at Brown for such former Göttingers as Lewy, Neugebauer, and Feller as well as for Willy Prager, who had been a *Privatdozent* for applied mechanics in Göttingen from 1929 to 1933.

Richardson was also a key member of a group of mathematicians who had shown the quality of leadership by their participation in the activities of the American Mathematical Society. With extraordinary energy and ingenuity these men and their predecessors had organized the mathematical society, solicited members, obtained support from business and industry, collected libraries, published journals, prepared the necessary equipment for mathematics in America so that it was already there when the refugee mathematicians from Europe began to arrive in 1933.

After graduating from college in Nova Scotia and serving as a high school principal for several years, Richardson had taken a second A.B., an M.A. and a Ph.D. in mathematics at Yale. He had then gone abroad for additional study. He had arrived in Göttingen in the year 1908 just after Courant, ten years younger than he and still a student, had become Hilbert's assistant. During that year Richardson had given Courant a mathematical paper to pass on to Hilbert for publication in the *Mathematische Annalen*. No one I spoke to really knew what happened. Courant said only, "I was very young and careless." Richardson's paper never appeared in the *Annalen*. In 1931, reviewing the second edition of the first volume of Courant-Hilbert, Tamarkin, a colleague of Richardson's at Brown, commented pointedly: "The bibliographical references are a little more complete in the present edition [1930] than in the first one. In this connection the reference to an unpublished paper by R. G. D. Richardson should be welcomed. . . . It contained numerous points of contact with the results of Chapter VI. . . ."

When Courant came to America, he tried immediately to remove the "dissonance" in his relationship with Richardson. On the surface he was more or less successful. The two men exchanged polite letters about the placement of refugees and the need for applied mathematics. But in 1941 the feeling of each that he was the champion of this subject so long neglected in America brought them into collision.

Since the outbreak of war in Europe, there had been increased interest in the applications of mathematics on the American side of the Atlantic. In September 1940, at a joint meeting of the American Mathematical Society and the Mathematical Association of America, Marston Morse, as chairman of the War Preparedness Committee, had recommended "that graduate schools extend their courses in applied mathematics . . . and that advanced students be urged to become highly qualified in one or more fields of applied mathematics." By 1941 plans were being made by the Office of Education for a defense-oriented summer program. Courant thought the proper place for such a program was New York City, told Richardson so, and began to make arrangements at NYU. Richardson went ahead with his own plans for setting up a program at Brown. In a short time he was able to make a public announcement of the creation at that university of "the nation's first center where engineers, mathematicians, technicians, and other specialists in defense production can devote their full time intensively to problems of higher mathematics as applied to industry."

The program at Brown in the summer of 1941 was "a very high level affair," according to Friedrichs, who taught a portion of one of the courses. Most of the faculty had come originally from Europe. Richardson was especially proud that he had obtained the advisory and instructional services of von Mises, the former director of the Institute for Applied Mathematics at the University of Berlin, who had recently emigrated to the United States from Turkey.

That same summer Courant circulated his proposal for a national science institute, and he sent Dean Richardson a copy. Richardson responded politely, saying that although he thought that such an institute was an important idea, he was thoroughly convinced that it should be attached to some existing institution. Still he was "vitally interested . . . ready to give time and thought to the advancement of the cause, whatever is finally chosen as a plan."

Shortly after receiving this letter from Richardson, Courant, chatting in the office of a friend, perceived lying on the desk among some other papers a letter which he recognized, although it was upside down, as coming from the dean at Brown.

"We had a long talk, this man and I, in the course of which we shuffled some papers," Courant recalled. "Afterwards I went on a little trip to the Adirondacks and when I unpacked my briefcase there were some papers in it which I didn't know. So I had by mistake swiped some of this man's papers, and there was a document by Richardson which discussed this proposal of mine and was really very hostile. He was very much against it —not against it, but against me doing it. He didn't think that foreigners should come in and make such proposals."

"Would that letter from Richardson be in the files at the institute?" I asked Courant.

"Oh no," he said. "You see, I stole it."

By summer 1941 most of *What Is Mathematics?* was in type, and Courant was preparing to write the preface. He had produced what he felt was a different kind of popular book on mathematics. The 1930's had seen a number of such books in English; but, in Courant's opinion, all of them had a serious flaw. The basic premise on which they were written was wrong. The understanding of mathematics could not be transmitted by painless entertainment any more than the appreciation of music could be conveyed by even the most brilliant journalism to those who had never listened intently to music. To understand mathematics, one must *do* mathematics. In his book he proposed to give his reader "actual contact with the *content* of living mathematics."

What Is Mathematics? presupposed only knowledge that a good high school mathematics course could impart. Technicalities had been avoided, and also emphasis on routine and "forbidding dogmatism which refuses to disclose motive or goal and which is an unfair obstacle to honest effort." On the other hand, the book was not "a concession to the dangerous tendency toward dodging all exertion." The promise which Courant held out to his readers was that in his book they would be able to proceed "on a straight road from the very elements to vantage points from which the substance and driving force of modern mathematics can be surveyed."

The title of the book still caused Courant some concern. It seemed "a little bit dishonest." At a party one night at Weyl's, he consulted Thomas Mann. Should the title be *What Is Mathematics?* or should it be something like *Mathematical Discussions of Basic Elementary Problems for the General Public*, which was more accurate but "a little bit boring"?

"Then Mann told me that he couldn't advise me, but he could tell me about his own experience," Courant said. "Among his German books published in English there was one—*Lotte in Weimar*—Goethe's Lotte—and shortly before the book was to be published Mr. Knopf came to him and said, 'Oh, now we should select a title; and my wife, who has a very good sense about these things, thinks we should call it *The Beloved Returns.*' Then Mann said he felt a little uneasy about this title—after all, the title *Lotte in Weimar* was just as good in English as it was in German. And Knopf said, 'All right—but I just want to make one remark. If it is published as *Lotte in Weimar*, then we will sell ten thousand or maybe twenty thousand copies; but if it's *The Beloved Returns*, we will sell one hundred thousand copies—and of course the royalties will correspond.' Mann said, 'I have decided—it will be *The Beloved Returns.*' "

Courant thanked Mann and went immediately to the telephone to call the printer.

Although young Robbins had been reading galley proofs and had made several trips to the printer in Baltimore, he had not yet seen a proof of the title page. Then, toward the beginning of August 1941, he saw for the first time *"What Is Mathematics? by Richard Courant."*

"When I saw the title page, I suddenly said, 'My god, the man's a crook. It was just a sort of a cold bath. What I regretted was not only that I was not going to get my name on the book—because I immediately knew that I was—[but that] it was the end of my feeling that Courant was a decent, honorable person who wished to foster young people's work without worrying about his own prestige, etc. . . . Later, of course, I heard over and over again 'Dirty Dick.' People found it very easy to believe this had happened because they had heard other stories. I had not heard any other stories. I knew nothing but good about Courant. . . . I had rather loved him."

Robbins consulted some of the other people at Washington Square about what he should do.

"And they said, 'Well, you have to understand that in Germany this has not been uncommon. Many books are purportedly written by some well-known professor but are in fact written by some young student of his as part of the educational process.' I said, 'Well, first of all, I am not a student of his; second of all, this isn't Germany; and third of all, I don't like it.' "

It is impossible at this date to determine the chronology of events not recorded in the correspondence between Courant and Robbins; but at some point, Robbins told me, he spoke to Hassler Whitney, under whom he had taken his Ph.D. at Harvard.

"When I told Whitney of this, he was highly indignant and he said, 'Well, you tell Courant that if he goes through with this, I will bring the matter up at the next meeting of the American Mathematical Society, and we will expel him from membership.' "

Robbins had "promised or perhaps threatened" as he reminded Courant in a letter dated August 17, 1941, to state his general sentiments concerning the wording of the title page in writing. He had postponed doing this for some time because his feelings on the subject were so intense that it seemed unlikely he and Courant could have a quiet discussion and "a heated argument might have had a bad effect on the book, or at least on its publication date." In the letter of August 17 he stated that although he recognized that the book was in all essential ways Courant's book, he himself had worked so hard on it and had become so emotionally involved with it that having his name on the title page equally with Courant's had become very important to him. Besides, he felt that while the practice might be different in Europe, this was the standard American custom in such cases. Everyone recognized that the first name given on the title page was that of the real author of the book and the second name that of the assisting younger colleague. As to the financial arrangements, he was perfectly willing to leave those entirely in Courant's hands. He simply asked that the title page read "by R. Courant and H. Robbins."

Neither Whitney's threat nor Robbins's determination to take legal action is mentioned or implied in Robbins's letter. It is Robbins's feeling that Courant learned of these from others.

At any rate, after receiving Robbins's letter, Courant agreed to change the title page.

Robbins explained to me that he wrote his letter of August 17 in the way he did to make the idea of coauthorship more palatable to Courant. It had that effect. When Courant showed me the letter, he told me that he felt that it expressed exactly the situation that had existed between himself and Robbins.

In the next few weeks of the autumn of 1941, however, Courant apparently became aware of the feeling on Robbins's part that he was in fact a coauthor. On September 28 he wrote a long, stern letter to the young man.

"I am under the impression that you have permitted yourself to drift or to be pushed into an awkward psychological position. I think it is imperative that you be fully aware of the actual situation. It is not a question how much time, energy, devotion you have spent on your collaboration. By conception, by planning, by content, by original mathematical ideas the book is my child and more than any other of my publications expresses my very personal views and aims. You are enough of an individual to have your own outspoken views that need not fully coincide with mine, and it would be only natural [for] some marked deviation . . . to develop in the future. For this reason I was and I am anxious that the matter of authorship should be clearly understood by everybody, in the first place by yourself. This is not a question of ambition, it is a matter of scientific responsibility. Naturally I depended on assistance. . . . Your cooperation far exceeded what I could expect from a competent mathematician with a high standing of his own. By no means do I wish to withhold praise and public recognition. When in your letter of August 17th, you insisted on having this recognition expressed on the title page, I consented immediately. Your letter reassured me that there was not and never could be an essential misunderstanding between us concerning the basic issue of authorship and that misleading impressions on the public were not intended. . . .''

What Is Mathematics?—by Richard Courant and Herbert Robbins—turned out to be much more successful than anyone, except perhaps Courant, ever expected it to be. As of the date of this writing, it has been translated into a number of languages and has sold well over 100,000 copies. It is often described admiringly as "a mathematical best seller." Robbins told me that he never received a regular accounting of the book's sales, but simply from time to time a personal check and a note from Courant saying, "Dear Robbins, Enclosed is a check representing your share of the sales of *What Is Mathematics?* for the year 19—."

He is, as he says, "heartily sick" of the whole matter. When people ask him who wrote *What Is Mathematics?*, he sometimes tells them, "Courant wrote the book, but he put my name on it so it would sell."

"The important thing," he told me, "is that it's a good book, and it's had a good effect."

Courant himself was always "a little bit disappointed" in the book; for in spite of its success it never reached to any appreciable degree that general public of "educated laymen," to whom he had hoped to be able to convey something of the beauty of mathematics.

By the time the academic year 1941–42 began, Courant had achieved one of his goals—*What Is Mathematics?* was in print—but he had not achieved his other goal of a national institute of basic science. He always believed that the opposition by Richardson was decisive. In retrospect, though, he said that he also found it "of course childish and unrealistic" that he had even thought he could carry off such a grand project at a time when he had not been in the country very long and had not understood the degree of resistance to government support of science which existed in the United States at that time.

"So did you give up the idea?" I asked.

"I didn't give it up," he objected. "I thought I would try other means."

TWENTY-THREE

I N SPITE OF seven years of effort, Courant had developed only a very modest operation at NYU by September 1941. A folder of that date announcing the establishment of an Institute for Applied Mathematics in connection with the Graduate Center lists an impressive number of regular and emergency courses; but when one examines them, the new "institute" turns out to be just another name for Courant, Friedrichs, and Stoker.

By Thanksgiving of that year, the German army had effectively occupied most of the continent of Europe. The Committee in Aid of Displaced German Scholars had long since substituted the word "Foreign" for "German" in its name. Since 1933 approximately 130 mathematicians, mathematical physicists, and statisticians had come from Europe to the United States. Not included in the total were the many European children who also came with their families during those years and grew up to be contributing members of the American mathematical community. One of this group is Peter Lax, the present director of the Courant Institute. As a 15-year-old boy, Lax was on the high seas with his parents on their way from Hungary to the United States on December 7, 1941.

In the United States on December 7, which was a Sunday, the broadcast of a concert by the New York Philharmonic was interrupted by the announcement that the Japanese had attacked Pearl Harbor. On Monday the United States declared war on Japan. By Thursday Germany and her European allies had declared war on the United States.

In spite of the fact that he had long expected and even hoped for the United States to enter the war, the actuality was traumatic for Courant, as it was for most refugees. In the midst of the abrupt adjustment, he wrote to Heinz Hopf in neutral Switzerland and asked him to arrange for "a beautiful bouquet" to be delivered in Göttingen on January 23—the card to carry the message that the flowers came to Hilbert on his eightieth birthday "from his American students and friends with all their love and respect."

After Pearl Harbor, Courant continually paced up and down the long corridor between the offices at the Judson Dormitory. When De Prima brought lecture notes to be stenciled in the office of the secretary, he would stop the young man and get him to pace back and forth with him.

"I am sure he would do that with other people too. He was always on edge. It was very noisy there on Washington Square South so you could just barely hear a little jingle when the phone rang. But he would always hear it and immediately jump over to his phone, in spite of the fact that there was a secretary there to answer it, and say, 'Courant speaking! Courant speaking!' And I came to the conclusion—although he never told me so—that he was just waiting to hear from people in Washington to whom he had offered his services."

As soon as the United States entered the war, Courant began to grab for the talent he was sure he would soon be needing. Eleazer Bromberg remembers that in the spring of 1942 he came to see Courant about completing an unfinished course at NYU. He had already visited Brown, where the long-established and influential Richardson had been able to develop a program in applied mathematics which was much bigger than the one at NYU. He had pretty well decided to go to Brown. "Why not here?" Courant demanded and immediately arranged a job for Bromberg as an instructor in the physics department.

As the United States began to give indications in early 1942 that it was preparing to strike back against the Japanese in the Pacific, Courant was quite delighted to receive a grant of $3500 from the Carnegie Foundation, which would enable him to support a few more people. He conveyed the news proudly to Flexner and others, including Warren Weaver.

Weaver had been one of the early appointees of Vannevar Bush's National Defense Research Council (NDRC), set up in June 1940. His first assignment—as the bombing of Britain began—had been to take charge of the "fire control" section of the NDRC, which concerned itself with devices and procedures to make certain that a projectile would hit the intended target. He had later moved with Bush to the Office of Scientific Research and Development (OSRD), which was created by executive order of President Roosevelt in June 1941 to coordinate defense research—as Bush said—"wherever it might be."

The emphasis on the design and production of hardware had tapered off even by 1942, according to Weaver, because in general new devices could not be conceived, designed, built, tested, improved, standardized, and put into service in time to affect the outcome of the war. On the other hand, the need for analytical studies had increased. By the summer of 1942, Weaver was recruiting mathematicians.

One of those whose services he obtained was Stoker, who with his student Bromberg was sent up to Columbia to join a group composed mainly of statisticians working directly with Weaver on anti-aircraft artillery problems. Stoker was impressed by the talent Weaver had assembled, but he was enormously unhappy at Columbia.

"How did you happen to leave NYU anyway?" I asked.

"Oh, it was one of the bargains Courant drove with Warren Weaver," the forthright Stoker replied. In return for Stoker, Weaver agreed to support Courant, who with Friedrichs was already doing work for the Navy on the transmission of sound under water. "There were just not very many applied mathematicians in the country at that time." This was no wonder, since a report issued in 1941 had concluded that American industry could not absorb more than ten applied mathematicians a year.

In the fall of 1942 Courant ran a seminar at NYU on the theory of stationary shocks, important in aerodynamics. He was particularly interested in the theory because of Riemann's pioneering work in it. That December he happened to run into John von Neumann in Washington. Von Neumann had been doing some work on moving shocks, important in explosion theory. Since he was about to leave for some new and secret project, he offered to have Raymond Seeger, a co-worker in the Navy's ordnance group for fundamental explosive research, brief the NYU people on his results. This was the beginning of extensive work at NYU on shock waves, mainly by Friedrichs. The early contact with Seeger, both while he was with the Navy and later when he was with the National Science Foundation, was to be of inestimable value to Courant and his enterprise.

By this time the OSRD had carried out a reorganization which had shifted the fire control problems to a new Division 7 and created another OSRD agency called the Applied Mathematics Panel. The panel was to be of general assistance in connection with the fast developing analytical and mathematical problems, not only for Division 7 but for all the other divisions of OSRD as well and, even more broadly, for all the military services and the war effort. Weaver was made its chief scientific officer. The call from Washington for which Courant had been waiting came very shortly. It was Weaver inviting him to become one of the half-dozen members of the new panel.

The Applied Mathematics Panel cut across all divisions of OSRD. It was planned as an organization of civilian mathematicians who would provide mathematical help to other scientists involved in military work or to parts of the military needing such help. It was hoped that by this means all the leading mathematicians in the country not already employed by the military would become fully involved in the war effort.

For his panel, which was officially known as the "Committee Advisory to the Scientific Officer," Weaver selected, in addition to Courant, T. C. Fry, L. M. Graves, Marston Morse, Oswald Veblen and Samuel S. Wilks. Years later, Weaver recalled the serious discussions which had preceded the extending of the invitation to Courant—"in view of the classified nature of many of [the] projects, whether it was prudent to include in the top governing committee, a man—however distinguished and able—who had been a member of the Imperial German Army during World War One."

At the first meeting of the panel, before Courant arrived, Fry of the Bell Laboratories said, "We must, from the very outset, make Courant realize that we view him, with no conceivable reservation, to be *one of us.*"

While Fry was saying this, there was a soft little knock on the door.

Fry, who like the others had always addressed him in the past as "Professor Courant," sought a way to express his friendly acceptance of another American.

"Come in, Dick!" he said.

Courant took naive delight in his new importance and influence as a member of the Applied Mathematics Panel. He bustled back and forth between New York and Washington, dropping well-known names not quite inaudibly, mumbling vaguely of important secrets, glancing constantly at his watch, complaining apologetically about "the almost inhuman burden of work under which we all suffer at this time."

He was constantly generating ideas. A stream of papers labeled "Memorandum to WW" began to flow from his desk.

Many of Courant's suggestions to Weaver had to do with personnel. One mathematician he immediately proposed for the panel was Griffith Evans of the University of California. He was also eager to bring in von Kármán "as a foremost representative of applied mathematics and applied science in this country," and a lengthy memorandum on the "psychological background" to be considered in approaching Kármán went to Weaver.

Evans did become a member of the Applied Mathematics Panel. Kármán, who was the scientific advisor to the Army Air Corps, indicated interest in the group; but he was prevented from joining by illness and the approaching end of the war.

Somewhat later Courant recommended to Weaver a young woman named Mina Rees, a junior mathematics professor at Hunter College, for the position of technical aide. This position was to turn out to be more important than originally expected, since for a time during the war Weaver was incapacitated by an ear infection which affected his sense of balance. Mina Rees then became in practice his representative on a day-to-day basis in relation to the panel and increasingly represented the panel in connection with its activities. The job was for her the beginning of an outstanding career in government and university administration.

"I really don't know how Courant happened to think of me," she said when I talked to her in her office in the handsome new building of the Graduate School of the City University of New York. "He hardly knew me except for mathematical meetings and things like that. I have always thought it may have been because a very good student of mine later went to NYU. Her name was Bella Manel—and it just happened through an accident of scheduling that almost all of her undergraduate work in mathematics was taken with me."

I told her Friedrichs had suggested to me that Courant had recommended her to Weaver simply "on the hunch" that she had great potential as an administrator.

"I think Friedrichs is probably right," she agreed after a moment's thought. "I think Courant had a real genius for identifying people's potential. He knew what you wanted, what you were really interested in, sometimes even before you did."

The Applied Mathematics Panel operated through "contracts" with various universities. The idea of supporting scientific work by means of such contracts was new in America—"and one of the great inventions which came out of the war," according to Mina Rees. A total of 194 studies were conducted. These were summarized in four published volumes after the war. In just one instance—the development of powerful new statistical techniques which improved the efficiency and lowered the cost of testing war matériel—more money was saved by the military in a few months than the panel had expended during it entire existence. The group working at NYU was only one of a large number of groups set up at institutions from coast to coast; but for Courant and his co-workers the period of connection with the Applied Mathematics Panel was more important than for most. It was the turning point in the long struggle to establish a scientific center at NYU.

With the prospect of an Applied Mathematics Panel contract which would provide a generous amount of money, Courant began to hire talent he had long coveted for his group. The first person he drew into the new work was Max Shiffman, who since obtaining his Ph.D. had been teaching mathematics at CCNY. Courant considered Shiffman the best student he had had in America—one of the best he had ever had—and he had long sought a way to bring him back to the Graduate Center. The second person he hired was J. K. L. MacDonald, whom he arranged to take on loan from the nearby Cooper Union. He saw MacDonald as having a mind "both analytical and inventive" and "embodying a unique combination of mathematics and theoretical and experimental physics and a keen sense of applications."

In addition to young people already connected with the Graduate Center, Courant brought in young mathematicians from other places. One of these was Bernard Friedman—"a first-rate research worker," in Courant's opinion, "of enormous versatility and power." Everybody at the Judson Dormitory recruited friends and acquaintances. By the end of the war the applied mathematics group there numbered more than thirty.

The first work of the AMP group at NYU was a continuation of special Navy assignments at the beginning of the war in the two different fields of underwater acoustics and explosion theory.

"Automatically," as Courant saw it, "[the work] developed on a rather broad front. For example, the whole field of interferences of nonlinear waves was studied, the phenomena of the problem of gas motion after underwater explosion explored, its connection to surface phenomena studied with the consequence of progress in the air-water entry problem. The explosion theory led to general studies in supersonic gas dynamics which [then] drew the group into work on jets, rockets, and jet propulsion in general."

What pleased Courant most was that in this way "a state of affairs was reached where knowledge and experience of the group as a team [were] utilized for very different agencies." This couldn't have happened, he maintained, if the group had been scattered or divided into sub-groups, each dealing with a circumscribed assignment. The *group* was important.

During the war, at NYU, as at other American colleges and universities, the number of full-time graduate students—there were only a very few even in normal times—declined drastically. There were, however, some part-time students who worked during the day in New York in government or defense-related industry.

In spite of the load of increasing war research and the few students, Courant insisted on continuing to teach. As the war continued, his lectures became less and less prepared. Finally they were not prepared at all. Notations would be changed ten or twelve times in a single lecture. At one point he was teaching a class in the calculus of variations on Tuesday nights and a class on partial differential equations on Thursday nights. The two subjects being related, when he found that he could not prove a certain theorem in the Tuesday class, he would mumble, "Ja, ja, well, I will prove that theorem in my Thursday class." Two nights later, when the necessity for proving the theorem again came up, he would tell the students, "Ja, ja, well, I will prove that theorem in my Tuesday night class."

"His were the most incredibly diffuse and complicated lectures I have ever heard," said Harold Grad, now director of the Magneto-Fluid Dynamics Division of the Courant Institute, then one of the few full-time students. "But beautiful. It is hard to describe why. But you learned. You always learned something that was more interesting maybe than what he had originally planned to teach. I always enjoyed listening to him."

The war period was an exciting time, remembered with nostalgia by everyone I interviewed who had been at the Judson Dormitory then. Suddenly, after more than seven years, the tiny enterprise on the second floor began to expand. Suite by suite the girls were pushed out. Finally the mathematicians had three floors of the four-story building.

Even during the war Courant continued to feel that scientific writing should not be abandoned. One of the most important contributions the group made to the war effort was the writing of a manual by Courant and Friedrichs on shock waves. The editorial assistant for this project was a bearded young man with a degree in English with whom Courant had scraped up on acquaintance while sharing a bench in Washington Square.

Courant also continued to think about the little book he had long wanted to write on Dirichlet's principle, but there was so little time—that would have to wait.

There was suddenly a fresh demand for Courant-Hilbert from scientists working with the military. Since all German properties in the United States, including copyrights, were subject to seizure by the Alien Property Custodian, the two-volume book could be photographically reproduced by an American publishing company under a license obtained from the custodian. In 1943 Interscience Publishers issued a photo-offset edition. Even though it was in German, it sold 7,000 copies.

In Germany—Courant later learned—a request by the research director of the Luftwaffe for a reprinting of Courant-Hilbert had at first been rejected on the grounds that one of the authors was Jewish. It had finally been granted—two months before the end of the war—with the stipulation that the run be limited to 500 copies for official use only.

The same year that Courant-Hilbert was published in the United States, Hilbert died in Göttingen, a few weeks after his eighty-first birthday. It fell to Hermann Weyl to evaluate Hilbert's life and work for that part of the world which was at war with Germany.

"It was appropriate that Weyl should do it," Friedrichs said. "Hermann Weyl was the mathematical son of Hilbert—he understood Hilbert as a mathematician much better than Courant did. It's true that Courant always considered himself the son of Hilbert—and he always played down what he owed to Felix Klein—but in fact he was the son of Klein. He learned from Hilbert, and he was greatly influenced by Hilbert. But he always took what he got from Hilbert and transformed it in the spirit of Klein. In a way it was that combination—Hilbert in the spirit of Klein—which was *Courant.*"

As pressure on technological centers to provide numerical solutions of partial differential equations increased, there was also a call for another old piece of work from Göttingen.

At Los Alamos, where Donald Flanders was heading the computing department, an ingenious system of punch cards was being utilized in order to solve—by means of finite-difference schemes—the hyperbolic partial differential equations for fluid flows that build up in the course of time. At first, inexplicable errors appeared in the results. When the mathematicians turned to the old Courant-Friedrichs-Lewy paper on finite-difference equations, they found there the explanation of their difficulties in an observation made by Lewy. Seizing upon the fact, which he had learned in

Courant's course, that disturbances of the solutions of hyperbolic equations, produced by initial disturbances, always travel with a definite finite speed, Lewy had recognized that the solution of such an equation cannot be approximated by that of the finite difference equation if the propagation speed for the latter equation is less than that for the differential equation. He accordingly had drawn the conclusion that severe restrictions must be placed on the choice of the mesh-width of the lattice in designing a scheme for obtaining the approximate solution of the hyperbolic partial differential equation by means of the finite difference equation.

"Once it is said, it is clear to anybody," Friedrichs said of the point made by Lewy, "but Lewy was the first to say it!"

At NYU the computation required for AMP contracts was organized into something of an operation by Eugene Isaacson, a former student who had been working for the Bureau of Standards since he took his master's degree. There, as part of a group which computed tables for military research, he had begun for the first time to see mathematics as a possible career. He had thought vaguely of going back to school after the war. A friend from the group at NYU had encouraged him to go and see Courant about a job.

"I remember almost too clearly," Isaacson told me with a rueful shake of his head, "I had to come back and see Courant on about five or ten different occasions, because he would see me and mumble something and it was never quite clear to me whether I had a job or not. He would always be fumbling with some papers while he talked to me. He had the property of never saying 'yes' or even 'maybe yes' unless he was absolutely sure that he could fulfill the commitment. So it was only after he had seen me about ten times that he finally told me that I would have a job here."

Isaacson's job was to set up a system for numerical calculation similar to the one under which he had worked at the Bureau of Standards. Today, in the age of electronic computing machines, it sounds trivial; however, it permitted a few experienced people with hand-operated calculators to supervise the work of a large number of less experienced people, who performed the bulk of the calculations with paper and pencil according to step-by-step mimeographed instructions which contained built-in checks of their work. To Isaacson's surprise Nina Courant volunteered her services to his group. He found the daughter of Carl Runge "extremely intelligent in how she looked at the subject" and very interested in learning some of the theory behind what she was doing.

While Isaacson was running the computing department, "learning a bit from the scientists involved about where the problems originated," he began to attend "classes." These were seminars or lectures given by some members of the group for other members. Everybody was a student, and often a teacher as well.

For Courant this maintenance of training during the war was an important principle.

"I *fought* for that," he told me proudly. "I opposed Mr. Bush, who was the head of the OSRD. He said I should not waste time and energy teaching when I could be doing research work for these war purposes, but I opposed him. I said it was self-defeating. So anyhow I organized it so that the people we hired for research work here should also be members of seminars, give courses and seminars themselves. I think that was really one of the most constructive things I did, that I resisted the idea that the Applied Mathematics Panel should only solve problems. I thought the purpose of the university, the teaching, the seminars should not be reduced. On the contrary, maintaining them would be very important for the future."

Toward the beginning of 1944, Courant became acquainted with E. S. Roberts, the chief engineer of the Chemical Construction Company, a subsidiary of American Cyanamid. Roberts, who was interested in rockets and intrigued by newspaper reports of "secret weapons" in Germany, played an active role in stimulating American research on jet propulsion during the war. He suggested to the NYU group that they investigate the flow through the nozzle of a particular kind of rocket. The result was a series of studies of flow problems in rockets and related structures and the incidental development of a computational procedure which turned out later to be very useful.

Courant was tremendously enthusiastic about Roberts, "a very dynamic, superbly intelligent and well-informed and inventive man, . . . one of those engineers that are not interested in money but in achievement and social progress." Friedrichs remembers how he and Courant and Roberts sat late in Courant's living room in New Rochelle plotting the future of the NYU group and of science in America. They were especially concerned about the continuation of government support of science after the end of the war.

"What we were really talking about," Friedrichs says now, "was something like a national science foundation."

The war work at "Courant's shop" (as Mina Rees called it) was carried on in a style somewhat different from that at other AMP contract institutions; Courant—even during the war—insisted on maintaining the basic character of the research program there. This was in opposition to the general attitude in Washington, Mina Rees told me.

"We—the Americans—were pretty much committed, in contrast to the English, to the notion that we would work on problems of immediate use rather than on basic research; and almost all the applied mathematics contracts were let out on that basis. Courant was impatient with the idea and insisted that he had to let his people and himself work on basic materials. So that Friedrichs's work, for example, was heavily on the development of the whole shock wave and fluid mechanics theory.

"But," she pointed out, "Friedrichs was indispensable in using the insights he got out of this basic stuff when people had practical questions. I remember one time when he and I went out to Caltech because they were having trouble—they had a rocket contract out there and there was something the matter—the rockets didn't take off properly. Because of his understanding of the flow mechanism, Friedrichs was invaluable in making suggestions. . . . Yes," she said in answer to my question, "they did finally get the rockets to work as a result of his help."

"Well, it wasn't quite like that," Friedrichs smiled. "I didn't ever actually solve their problems for them. Maybe I acted as a kind of catalyzer. Since I did not know too much about the engineering aspects, I had to ask them quite a number of questions. Naturally my way of formulating questions as a mathematician was quite different from what they were used to. This forced them to look at their problems in a different, more fundamental way. That probably helped them. In the end, of course, they solved their problems themselves."

Friedrichs's remarks illuminate the way in which for the most part the applied mathematics group at NYU provided assistance for the military and for "real applied mathematicians"—i.e., engineers. Although the group dealt with such practical-sounding subjects as explosions, detection of underwater sound, flow of air through jet nozzles, design of ramjets, supersonic flow and shock waves, they usually treated these subjects from a rather theoretical point of view.

There were cases in which this approach was directly and practically very effective. But for the most part the people with the problems read the NYU reports on the mathematical theory involved—often in the so-called "ideal" or "generalized" case, which was not at all the same as the specific case they were dealing with—and as a result they were able to recognize the mathematical situation which concerned them and—as Friedrichs said —to solve their own problems.

By the beginning of 1944, it was clear that an Allied victory was inevitable, but on both sides of the world the war went on.

"Unfortunately, I cannot be so optimistic as to count on a foreseeable end. . . ," Courant wrote in January 1944 to Flanders at Los Alamos.

Six months later, on June 6, 1944, the Allied Forces landed on the beaches of Normandy. By the middle of August the west flank of the German army had collapsed.

Courant wrote again to Flanders:

"The time has come for thinking of post-war plans."

TWENTY-FOUR

ON MARCH 23, 1945, the United States Ninth Army and the British Second Army crossed the Rhine, one half of a pincer operation planned to trap the industrial heart of Germany. On April 8, as they swept eastward, Göttingen fell in the path of the Americans. "There was one unpleasant half hour during which we were attacked by artillery," Franz Rellich, who was living with Herglotz at the time, wrote to Courant. In Herglotz's garden a shell exploded, broke a window, and deposited an apple tree on the balcony of the second floor. "Ten minutes later American soldiers were in our house and asking for water. The Thousand-Year-Reich was over!"

Four days later, in the United States, Roosevelt died. He had been president in 1933 when the first refugees from Hitler had begun to arrive, and they had never known another president.

Fighting still went on in the Pacific, but everybody at top policy levels in Washington agreed that the Applied Mathematics Panel should be phased out as soon as the war was over. It was intended, however, that the practice of government encouragement of scientific research by contract would continue under the newly organized Research Board for National Security (RBNS). Courant continued to push the ideas he had advocated while a member of the Applied Mathematics Panel. The new group should "continue and expand work of *a basic character* . . . the wisest policy," he maintained, "would seem to be the most generous." It should also tackle the problem of training of personnel—otherwise "no financial support, no panel, and no organization [will be able to] achieve the desirable results." Utmost flexibility should be one of the guiding principles. "No one has sufficient wisdom to determine . . . what fields ought to be attacked by scientists."

In the year following the European victory, the terrible story of what had happened to relatives, friends, and former students in Germany gradually began to be pieced together from letters and messages.

Of Courant's aunts and uncles, six had been alive when the war began. Three of these had emigrated, one had managed to live in Berlin throughout the war, and two had died in Theresienstadt, the "model concentration camp" for the elderly and the well known. Nineteen of Courant's cousins had left Germany to settle on five different continents. Four had survived with their parents in Berlin. Two had committed suicide. Five had died in the gas chambers.

Among the established mathematicians in Germany, there had been several suicides but almost no deaths directly at the hands of the Nazis. Nearly everyone who had been in danger had been got out, in some way, before the war began.

Courant was relieved to hear that Rellich, who had been a professor in Dresden at the time of the fire-bombing of that city, was safe.

"Dear Courant," Rellich had written ten days after the American occupation of Göttingen. "Since in the later part of this letter I will talk about myself all the time, I will at the beginning at least express the hope that you and your family are well and that the same is true of the larger family which consists of Friedrichs and Lewy and Neugebauer and Busemann and Feller, etc., etc.

"The following happened to me during the war:

"Since the summer of 1939 I have been in Dresden . . . and the great number of students there made it possible that I was not inducted into the army and also not completely submerged in 'war mathematics' as were most of the colleagues of my age. Since '41 I have had a proper apartment with four rooms, an amazing achievement for a bachelor; but this Dresden idyll came to an end with the air raid on the 13th and 14th of February 1945. Since I was staying with a colleague that night, everything of mine was destroyed. . . . I have not a single one of my papers, and actually I no longer know who I am.

"Anyway I myself did not burn, and this alone is (for me) quite pleasant. More pleasant things are to come. I received an order to continue my 'war-important' mathematical researches in Göttingen, and so here I sailed into the Fourth Reich. . . .

"There have been many moves to Göttingen, the trend from east to west is quite apparent . . . and these have resulted in an abundance of mathematicians here—which almost reminds me of times past, at least as far as quantity is concerned. The Mathematics Institute, however, sleeps like Sleeping Beauty and obviously awaits the kiss of the Prince. It would be wonderful if you felt like playing the role of the Prince."

There were many details in the letter of colleagues, deaths, illnesses, "existence."

"Of myself," Rellich concluded, "I can say very little at the moment except *sum*. In many ways this condition of being unburdened is comparable to that of *anno* 1926 when I first arrived in Göttingen.

"Only this time I miss you very much."

Courant would have liked to go to Göttingen immediately to see what he could do to help Rellich, other old friends, and members of Nina's family; but the transition from war to peace in the United States, the closing down of the Applied Mathematics Panel, the reorganization of the government's relation to science, the new possibilities in obtaining personnel—all these required his presence. Yet the problem of the rehabilitation of Germany occupied a great deal of his thought.

He got a letter on this subject from Harald Bohr in July 1945. It was his first direct communication from Bohr since 1943 when the two Bohrs—Jewish on their mother's side—had found it necessary to flee Nazi-occupied Denmark for neutral Sweden. Since Niels Bohr, under the name of Mr. Baker, had been almost immediately spirited away to the United States to assist in what was then referred to as the Manhattan Project, Harald Bohr had felt it "a kind of duty" to remain in Sweden even though arrangements had been made by Courant and others through the Rockefeller Foundation for him also to leave.

"During the time in Sweden," Harald Bohr wrote to Courant in English, "I often thought of writing a real letter to you, and in a way I cannot quite understand why I did not do it, but everything was so abnormal and absurd that in fact I did not do many things which in themselves would seem most natural and obvious."

Bohr too was concerned with the problem of Germany.

"To put it briefly, I should say that in my opinion it is both necessary and just to try to distinguish strongly between the notions of 'German' and 'Nazi.' Or to express it another way: The belief very common, for instance, in this country, that Nazism is something deeply and fundamentally fitted for human beings of German origin, seems to me very dangerous in two respects. On the one side it gives a quite absurd credit to the cruel and inhuman Nazi-movement to consider it as something 'natural' for a rather great proportion of mankind, and on the other side it gives a base for considering a possible reduction [sic] of the German people as on beforehand hopeless."

Although Courant agreed with Bohr in principle, he found himself repelled by the tone of almost all the letters which began to come to him from Germany. They were querulous and critical of the Allies. They conveyed no sense of responsibility for what had happened to the Jews or for the holocaust that had been brought upon the world by Hitler. Letters like Rellich's—uncomplaining and full of hopes and plans for the future—were rare. Courant wrote to his former student, "I cannot tell you how much I admire your attitude!"

As soon as the war ended in Europe, young scientists who had not yet finished their education began to think of returning to school. In summer 1945 Louis Nirenberg, who was working in a physics laboratory in Montreal with Courant's daughter-in-law Sara, asked her to get him some advice from her famous father-in-law, "just a suggestion as to where to go to study physics in the United States." She returned with an offer of an assistantship in mathematics at NYU. It was Courant's idea that the young Canadian spend a year or so learning mathematics, get his master's degree in that subject, and then go on and get his doctor's degree in theoretical physics.

Although Nirenberg had never heard of NYU, he accepted Courant's offer. Today he is a very pure mathematician.

"I haven't done any work at all in connection with physics," he told me. "I think people who turn from physics to mathematics are drawn to it because of its more abstract nature, and so they tend to go all the way."

Like most of his German-born colleagues, Courant differed from American mathematicians of his age in that he had personally experienced combat in the First World War. This fact enabled him to empathize with the men still fighting in the Pacific but also made him determined to save gifted young scientists from their fate. In 1945, when the Hungarian-born Peter Lax was drafted, Courant hastened to pull a few wires; and Lax very shortly received orders to report at Los Alamos.

For Lax, the Manhattan Project was a fascinating and formative experience. Later, after he got his Ph.D. at NYU in 1949, he went back to Los Alamos for a year and then, after that, for several summers as a consultant.

"That's really how I got into applied mathematics," Lax told me. "It's true I picked up a lot at NYU, but it made a crucial difference to be associated with a project that had some very definite technical goals. On the basis of that experience, I would say that's the way to learn applied mathematics. It's not just an armchair business."

Courant and many people in his group suspected, although none of them knew for sure, what was going on at Los Alamos. Stoker still remembers how, in summer 1945, running into Robert Oppenheimer on a train going west and being somewhat annoyed by Oppenheimer's condescending attitude toward the novel he was reading, he snapped: "You know, what you're trying to do out there, you are never going to be able to do it!"

Shortly afterwards, the first atomic bomb was dropped on Hiroshima. On August 14, 1945, the Japanese accepted the terms of the Potsdam Proclamation—the total surrender of all military forces.

The war was over.

On September 17, 1945, Göttingen reopened its doors, the first German university to do so. A second letter to Courant from Rellich, dated September 26, described the event.

Allied lawyers, busy interrogating the population, were using most of the rooms in the Mathematics Institute. A few days before, they had been instrumental in having Hasse dismissed from the faculty.

"This did not take place, as it seems, on the strength of the questionnaire of the military government but rather following some conversations Hasse had with the Americans. Herglotz and Kaluza sent a letter to the rector protesting the dismissal and pointing out the great mathematical achievements of Hasse, but without success."

When I talked to Hasse some thirty years later, he told me that he himself did not know whether he was dismissed on the basis of remarks he had made to two Americans who had come to his home to obtain a manuscript concerning his wartime investigations of ballistic problems—or on the basis of some things he had said in the first faculty meeting:

"My political feelings have never been National-Socialistic but rather 'national' in the sense of the Deutschenationale Partei, which succeeded the Conservative Party of the Second Empire (under Wilhelm II). I had strong feelings for Germany as it was created by Bismarck in 1871. When this was heavily damaged by the Treaty of Versailles in 1919, I resented that very much. I approved with all my heart and soul Hitler's endeavors to remove the injustices done to Germany in that treaty. It was from this truly national standpoint that I reacted when the Faculty more or less suggested that such a view was not permissible in one of its members. It was also the background for my remarks to the Americans. They were talking about reeducating Germany, and I said some strong things against this. It irked me that everything against Hitler was desirable, and everything that he had done was wrong. I continued to be a national German, and I resented Germany being trampled under the feet of foreign nations."

That first post-war September Göttingen was very lively. Rellich, lecturing in the place of Siegel, had more than three hundred students attending his calculus lectures. But the return of Siegel would be decisive for Göttingen, Rellich told Courant. "More important, it would give some hope again to all mathematicians in Germany, a hope that not everything is lost here forever."

The happenings he was reporting in his letter were, he wrote, "just the first flowers on a grave."

"Unfortunately it is not only thoughts of the past that are depressing. It is terrible to think of the fate of my friends in Dresden and Leipzig, many of whom will have starved to death in another six months."

At NYU, that same September, Courant faced a new problem in relation to the group he had built up during the war. There was a general movement to continue government support of science in the universities and a feeling on the side of the universities, which had in the past been suspicious of such aid, that some sort of support from the federal government was going to be necessary to keep American institutions on an equal footing with the state-supported institutions of Europe and Russia. It seemed certain that a bill establishing a national science foundation would be enacted and would make sizable sums available for scientific research. At the same time, closely connected with this development, efforts were being made to secure the continuation of some specifically military research with the gap filled by the Navy, which needed science more than the Army. At the moment, and in the foreseeable future, it was going to be easy to obtain money under contracts from industry and from the military; but such a situation was not going to last forever—and in the long run, in Courant's opinion, it was not desirable.

The numerous memorandums which he had composed in 1940 and 1941 for "a national institute"—"a basic science institute"—"an emergency institute"—lay in his files. He was still convinced that America had need for a high level scientific facility combining research and teaching in some central location—or even perhaps, as an alternative, a group of smaller institutes in different parts of the country.

Now, at the end of war, he thought he saw a different way to approach this goal, one which would solve the problem on a personal as well as on a national level. The type of regional center he had in mind could be put into operation immediately at New York University—"a natural development," as he put it, "of consistent though unobtrusive efforts" which had been being made for a decade.

Asking Warren Weaver for names of people to approach for the necessary outside financial support, he wrote a little wearily: "After many disappointments it may seem foolish for me not to give up and rather concentrate on my own research work. However, experience during the war and observation of the present trends have further strengthened my feeling of responsibility to go on with my efforts even if they have only a very limited objective."

In spite of Courant's fears, he did not lose any of his key personnel during the year 1945–46. Support for the group was beginning to come increasingly from Navy contracts.

"Our work now is very interesting and we need not compromise our scientific conscience if we go along the same lines," Courant wrote to Flanders. "Nevertheless, we are all gradually trying to take up mathematics for its own sake."

Ultimately, the Navy's interest in supporting research led to the establishment of the Office of Naval Research (ONR). Courant and his people were drawn into the new group by their friend Raymond Seeger.

Stoker told me that he has always felt that Courant deserved a great deal of credit for the post-war intervention of the government in research, particularly that of the defense departments.

"That started with the ONR; and we had, I believe, the first such contract in mathematics. The way that was formulated set the pattern for all the others afterwards."

Since the war Courant had heard very little about what had happened to his old friend and publisher, Ferdinand Springer. Then, on a train in November 1945, he picked up a copy of the *New Yorker* and found in the middle of the "Report from Berlin" the following sentences:

"Not long ago, I had the luck to talk with a German who, just after the fall of Berlin, was permitted by the Russians to look not only at a corpse that some Germans think may possibly be Hitler's but also at one that was undeniably [that of Goebbels].'' The German to whom the *New Yorker* correspondent had spoken turned out to have been Ferdinand Springer.

The report went on to say that in late February 1945, having been evicted from his firm by the Nazis, Springer and his family were living with friends in Pomerania. When the Red Army arrived on its way to Berlin, he was taken into custody by the Russian secret service, the NKVD, and asked to list all journals and books he had published. He was hardly into the onerous task when the major questioning him suddenly demanded to know if he had published a genetics journal. Springer replied, "Ja, *Der Züchter*.'' The major, a professor of animal husbandry in a Siberian university, beamed: "*I* was one of your contributors!'' It was, as the *New Yorker* correspondent commented, an example of the best in publisher-writer relations. From that moment on, Springer was treated as a friend. He remained with the major and the NKVD, which followed directly behind the front, and entered Berlin under the fire of German machine guns.

The first letter which Springer sent abroad went to Courant on April 11, 1946. By then the 65-year-old publisher was able to report a certain amount of progress in "rebuilding.'' He planned to decentralize his firm, not only on account of the four zones into which Germany had been divided by the Allies, but also because of his expectation that in post-war Germany culture itself would be decentralized. His first publishing ventures would be made in Heidelberg, in the American zone, and Göttingen, by now under the British. Mathematics had first place in his publishing plans.

Although Courant could not say how much help Springer could expect from other former authors now in the United States, he wrote that he himself would be happy to give him the publication rights to *What Is Mathematics?*

In his own operation at NYU during this period, Courant was making a very great effort "to stabilize" by obtaining academic appointments for the people who were working under the contract with the Navy. He was also eager that Flanders, still at Los Alamos, should return to the faculty. The secret wartime development of the electronic computer had been made public at the beginning of 1946, and Courant hoped that Flanders would take over a more advanced computing program at NYU.

"All the questions related to the big new fashion of machines, etc., seem now to be much more interesting and fascinating from a mathematical angle than I originally thought," he wrote to Flanders.

But Flanders was hesitant about returning.

"I hate to refer to my inferiority complex again, but I think that is the real basis of my interest in staying here. At the university I am constantly oppressed by the feeling that I have gotten myself into a situation with which I am unable to cope, and I count the years until I can retire, hoping that at least I can get my children educated before I or the situation crack."

Courant hastened to respond.

"Of course, we are all looking forward to the day when you will come back . . . , [but] any ideas of leaving us in the lurch . . . are out of place. . . . Our working philosophy has been, and is becoming more strongly so, that there should not be job assignments to be attended to by whomever one can find to fill the demand but rather that the existing human beings and teams are the primary element and projects and assignments should be adjusted to them."

After a visit from Courant in Albuquerque and a long weekend by himself in Berkeley, Flanders finally wrote that he had made up his mind to return to New York.

Chancellor Chase, under Courant's prodding, had earlier appointed a committee to evaluate Courant's idea of establishing an institute for advanced training in mathematics and mechanics at New York University. The chairman—at Courant's suggestion—had been E. S. Roberts, whose wide interests in science and education continued to impress Courant and seemed to him to more than make up for the fact that Roberts lacked the contacts of someone like Rear Admiral Lewis L. Strauss, whom he also considered as a possibility for the chairmanship.

By the end of the academic year 1945–46, the evaluating committee had submitted its report. Although recognizing much in the NYU situation that would not be favorable for the kind of advanced scientific institute Courant proposed, the committee recommended that the plan be put immediately into operation:

"The program envisages research integrated to a much higher degree with teaching than is ordinarily done in American graduate schools. In this respect, as well as in emphasis on the connection between mathematics and applications, and in emphasis on teamwork, [it] differs distinctly from the average graduate school program. It is hoped that the development of the New York University Group will stimulate similar developments in other graduate schools.''

In spite of this encouraging conclusion, the university took no immediate steps toward establishing the proposed institute.

It is impossible to convey in a linear narrative the multi-dimensional nature of Courant's activity in the years following the Second World War, and it is not actually relevant to this narrative. Still, no one who lived through that period with the group at NYU can separate even the larger events of the time from Courant. He seemed to them to be involved in everything that related in any way to the scientific community. In July 1946, when the first peace-time test of an atomic bomb was carried out near Bikini, Courant was among the scientific observers.

Later that same summer he went to England on a scientific mission connected with the Office of Naval Research. There was a possibility that he might get to Germany, but the press of duty and the lack of time prevented his going. When he returned to New York, he found that what he had feared had begun to happen. His group was breaking up. While he had been in England, De Prima had accepted a position at Caltech.

At the same time Courant lost De Prima, however, he gained Fritz John as a regular member of the NYU faculty. It was the first real addition he had been able to make to his enterprise since the hiring of Friedrichs and Stoker almost ten years earlier.

As soon as their wartime manual on shock waves was declassified, Courant and Friedrichs began to prepare an enlarged version for publication. In the fall of 1946, Courant hired Cathleen Morawetz, the daughter of the applied mathematician J. L. Synge, and put her in charge of collating sections that had been farmed out to various young people and of checking Courant's and Friedrichs's English.

She told me that at the time she—like Courant's daughter Gertrude— had recently married. Her father and Courant, running into each other at a mathematical meeting, had bemoaned the fact that now their daughters would not go on in their chosen fields of mathematics and biology, respectively.

"Ja ja, well, you can't do anything about my daughter," Courant had sighed, "but maybe I can do something about yours. You should send her to see me sometime."

Cathleen Morawetz is now a professor of applied mathematics at the

Courant Institute and a trustee of Princeton University as well as the mother of four grown children. When she came to work at NYU, she had no intention of continuing her studies—"After all, I was married!"—but Friedrichs was lecturing on topology that year and "everybody" was going to hear him and so she went too.

Courant's work on the shock-wave manual came to her in his scrawling, almost illegible handwriting or, sometimes, in typescript. She told me that he was the only person that she ever knew whose typing was characteristic. Letters were struck one on top of another, words were x'd out—typed in— and x'd out again—perhaps with a question mark. The question mark would be very light, because the key had been hit indecisively. Several words—all synonyms—would be typed in order, blanks would be left, or sometimes a word typed in but then enclosed in parentheses to indicate the tentative nature of its choice. In contrast, although Friedrichs wrote and rewrote constantly, he always gave her a very neatly copied manuscript. If anything was crossed out, it was thoroughly inked over; otherwise, Friedrichs knew, Courant would read what had been crossed out and not what had been left.

I asked her how such different people had been able to collaborate at all.

"The way it was," she recalled, "one or the other of them would take a section from the manual and rewrite it. If Courant did it, then it went to Friedrichs. And Friedrichs would look at it and grumble that it wasn't sufficiently exact. He would rewrite it, and it would become all *if's* and *but's*. Then Courant would take it, and he would mumble and groan that it was much too complicated. Then he would rewrite it. Then Friedrichs would take it back and say it wasn't precise enough. The process went on many, many times for each section. When it came back to Friedrichs, he would put in again some of what he had had before, but not so much. Then the next time Courant wouldn't take out so much. They were both pretty determined about what the end product should be, and they were both quite willing to do an awful lot of work. So they were really great at cooperating, but that is the way the cooperation took place. I never remember a single session where they both sat down together over the manuscript."

By June 1947 a year had passed since the evaluating committee's recommendation that an institute of mathematics and mechanics be established at NYU. With the exception of Fritz John, none of the younger people had been given academic appointments of any significance. All over the country the need for personnel was urgent. What Courant referred to as "attacks from the outside"—i.e., offers from other universities—were becoming more frequent and more difficult to ward off. Money was no problem. Government support of research seemed assured for years to come; but young men like Max Shiffman and Bernard Friedman, approaching or in their thirties, wanted academic appointments. The university administration hesitated to commit itself by hiring them. What if government support for science were to be suddenly withdrawn?

Stoker, especially, urged Courant to push the matter of proper support with the university, which was profiting greatly from the group's government work.

"Courant would go out and get money from the outside," Stoker told me, "but he would never ask the university for anything, unless he was pretty sure he was going to get it. Up until the time that I succeeded him as director, almost all of us—including Courant and Friedrichs and me—were paid, say, not more than one quarter of our salaries by the university. The rest of it came from many, many contracts. The university itself was absolutely no help to us in getting these or in carrying them out. It was a bitter fight all the time to get space or any of the things we needed. I found that ridiculous and used to urge Courant—'That's no way to do—it's out of line with what's done everywhere else—you don't improve your bargaining position by giving away to people all the time.' But Courant always preferred to take the most roundabout way instead of attacking things frontally. I used to tell him, 'If there are ten ways to do a thing, you will choose the one which is the least direct.'"

I wondered aloud why that was.

"Oh, that's just Courant, and just the opposite from me. In a way he liked that kind of intrigue, that was fun for him, he enjoyed it. And he always disliked making decisions. He would postpone decisions as long as he could and would even then always, if possible, fix things so that he would have a way to retreat. Even when he was quite sure he knew what he wanted. But his effects were very positive. That you have to say. Oh yes, he knew how to do things!"

In June 1947 the university did finally provide larger quarters for Courant's group, renting the entire second floor of the Bible House near the Astor Place subway stop. The first floor housed the Bible Society; and the third floor, the staff and equipment of the *New Masses*. Again, a new address was accompanied by a new name. Courant liked the concept of *an institute*—according to the dictionary definition, "a unit within a university organized for advanced instruction and research in a relatively narrow field"—and he always chose to use the word *institute* instead of *department* because, as he told me, "it permitted us to keep the options open for the future."

Before he left for Europe that summer—for his first visit to Göttingen in thirteen years—he again ordered a new letterhead:

NEW YORK UNIVERSITY
Institute for Mathematics and Mechanics

TWENTY-FIVE

COURANT'S trip to Germany in the summer of 1947 was financed by the Office of Naval Research. His assignment was to visit a number of universities and technical schools and to determine the extent of German progress during the war in the development of computing machines. In his new importance he was able to convince the ONR that he needed someone to assist him. It was arranged that Natascha Artin would accompany him on the trip.

The year before, when Artin had received an appointment at Princeton, Natascha had sought Courant's advice about finding a job in the east. He had responded by creating a position for her at Washington Square. She would sift through the mathematical reports that were constantly being received from the government and the military, and see that they were directed to those members of the staff to whom they would be of interest. Although she had no degree in mathematics, she had studied with a galaxy of famous mathematicians and physicists at Hamburg. These included—in addition to Artin—Blaschke, Hecke, Jordan, Pauli and Schreier. But she had married before getting her degree and had felt she should not go back to school to her husband's colleagues.

On June 14, 1947, she and Courant left New York for London, their journey sped along by Donald Flanders's good friend in the State Department, Alger Hiss. After a few days in England, where they found Hardy very ill but optimistic about the future of science in Germany, they flew to Frankfurt.

On June 20 they stepped out of their American plane onto German soil. In the journals which they kept of the journey, they both described how their eyes searched for the landmarks they had known. There were almost none. But the physical devastation of the city was much less affecting than the demoralization of the population. They were shaken by the sight of great crowds of ragged, hungry Germans, many of them begging.

During the next six weeks Courant and Natascha traveled extensively over the American and British zones of Germany, usually with a jeep and a driver, going as far north as Hamburg and as far south as Munich and, in Natascha's case, Vienna. They talked to innumerable people—former colleagues, former students, former friends in the non-academic world, relatives, military personnel, industrialists, young men and women, children, drivers, clerks, "a sad and complaining usherwoman" at a performance of "Theseus" in Göttingen.

Natascha was always to be amazed at Courant's ability to converse with people.

"He has a very good insight into the person he talks to," she told me. "He knows immediately what to talk about, and people become interested. It's certainly not his looks nor his superficial behavior. But there *is* something about him. I would still put a big question mark about what it is. I really don't know. There is no doubt that it exists. Because I've seen it!"

Warren Weaver had asked them to look for German scientists between the ages of twenty and twenty-five who might benefit from spending time in the United States. At the university in Darmstadt, the buildings of which had been almost completely destroyed, they found more than 2,000 students devoting half a day every two weeks to the construction of new buildings—a condition of their enrollment. However, the rector of the university—the physicist Vieweg, who in 1943 had publicly refused to join the Nazi Party—told them of great difficulties with the young people because of poor preparation and almost complete lack of ethical standards. For some, it seemed that the only thing the Nazis had done wrong was to lose the war. They began to feel that Warren Weaver would have to raise his age limits.

In Heidelberg they visited Ferdinand Springer. Courant noted "slight lack of resonance" at first. But he found Springer as optimistic and energetic as ever—forty-two journals were "appearing," Courant's calculus was being reissued. In Heidelberg they also visited the philosopher Karl Jaspers, who told them of the great need for contemporary books.

In Marburg they climbed the streets of the old town to see Kurt Reidemeister. Of all the people they talked to, he seemed physically the weakest; but he was full of plans for reconstruction.

"Reidemeister and his group are exponents of what we consider the proper attitude," Natascha wrote in her journal. "He is afraid of people leaving Germany permanently, but very much in favor of half-year or year fellowships. . . . [But he] thinks that there are no young people from 20-25 who would fall into the class of people [Warren Weaver asked us to look for] . . . . [These] are only now beginning to study because of the war. He thinks that people from 30-35 should be considered. Here the difficulty is that most of them are *belastet* [a German word meaning "subjected to a burden"—thus, carrying the stigma of having been National Socialist

Party members]—but he is sure that many of the *belastet* have not mentally been Nazis. [They] had to join the Party in order to keep their positions. . . . [Still] he said that 90 per cent of the population, including academic people, are dangerously but not hopelessly nationalistic. The natural science people in general much less, however."

From Marburg the visitors returned to Frankfurt and then flew to Munich. There and in the rest of the south they found much less enthusiasm for reconstruction than in the middle of Germany. Courant summarized many different conversations in his journal:

"Fear of Russians. Bitterness against French. Rumors also of American mismanagement. General lack of understanding for what America actually does to help the Germans. Little contact between scientists in different towns. None with abroad, almost none with Austria. . . . [Criticism] of German administration. Small-time politicians, no understanding for cultural issues. University has no support from them. Complaints also about zone competition. French do not permit some scientists to travel to other zones. Americans and British likewise compete for scientists and, allegedly, impose restrictions. . . . [Many scientists] do not dare travel through Russian zone for fear of kidnapping, which sounds unbelievable but is universally accepted as real danger."

After Munich the two visitors separated. Natascha went on to Vienna; and Courant, with some misgivings, proceeded by plane and car to Göttingen. Arriving there on the evening of July 3, 1947, he noted only, "Arrived, slightly nostalgic, in Göttingen about 7:30."

Physically the town was almost undamaged—it had been specifically exempted from Allied air raids, according to von Kármán. But people pressed upon him from all sides. There seemed to be thousands enrolled at the university. Lecture halls were often so crowded that students had to push their way in.

Courant found association with some of his old friends and colleagues very depressing: "Absolutely bitter, negative, accusing, discouraged, aggressive. Main point: Allies have substituted Stalin for Hitler, worse for bad. Russia looms as the inevitable danger."

In contrast was a pleasant evening spent with Heisenberg. He reported, "Heisenberg very superior, quiet, not complaining, and basically positive. Quite active scientifically, which of course is basis for psychological equilibrium." But another day, discussing politics, he found that Heisenberg "came out finally with the same stories and aggressiveness against Allied 'policy of starvation' [this referred to the dismantling of

German factories] as the less cool and more emotional people." Courant concluded that the physicist was "still in need of education, which I hope will be provided [when he visits] Niels Bohr in August."

The aerodynamics institute had become "a veritable fortress." Although ill and depressed, Prandtl was mentally active. He had given much thought to analog computing machines with a view to meteorological computations. The dimensions of the machine he was constructing had been determined by the size of ball bearings found by chance among war surplus.

By the time of Courant's visit, Rellich had become director of the Mathematics Institute. I got a picture of how Rellich operated in this position from Jürgen Moser, now a professor at the Courant Institute but at that time a 19 year old student recently arrived in Göttingen from the Russian zone.

After having crossed the east-west border under gunfire, Moser had presented a letter of recommendation to Rellich from his gymnasium teacher. Rellich had been immediately interested and friendly. He helped Moser find a place to live in the overcrowded town—"One could not register as a student without showing that he had a place to live, and he couldn't get a place to live unless he was a student." Afterwards, whenever he saw the youth, he greeted him in the Austrian fashion, "Grüss Gott, Herr Moser!"

"I was always amazed," Moser told me. "One is a student, one is rather shy, and to be addressed personally was somehow so—" He could not find the appropriate English word. "And then there was the matter of food. We students were literally starving, and Rellich managed somehow to get CARE packages and distribute food to us."

"Do you feel that Rellich was in a way following the example set by Courant?" I asked.

"I think so, but that's of course hindsight, because I knew Rellich first. I was always just amazed by him," Moser repeated. "I can illustrate with another, more scientifically interesting example:

"Most of the students who came to Göttingen then were of course much older than I. One of these was a man named Erhard Heinz—he is now a professor in Göttingen. He had been a student of Rellich's in Dresden, had been drafted, then captured and sent to England as a prisoner of war. Of course prisoners of war remained prisoners beyond the war, but all the time Rellich kept up a correspondence with Heinz and provided him with mathematical problems—otherwise one gets dull right away, you know. Later Heinz came to Göttingen and caught up very quickly. I remember him running around in these horrible clothes from the prison camp and living somewhere by himself, very lonely. Then, at Easter, Rellich sent him three colored and decorated Easter Eggs, each of them with a mathematical problem, an equation, on it. It was a charming thing to do, very typical of Rellich. I cannot imagine Heinz having developed the way he did without Rellich."

At Rellich's invitation Courant gave a lecture and soap-film demonstration on one occasion and a lecture on partial differential equations on another. Afterwards there was tea at the Mathematics Institute and a walk up the Hainberg for dinner, as in the old days.

Since the end of the war, Courant had tried persistently to find out from trusted friends like Rellich how various German mathematicians had behaved in relation to the Nazis. In Göttingen also, everywhere he went, even with Nina's family, the unspoken question lay below the surface of the conversation.

"I found very few people in Germany with whom an immediate natural contact was possible," he wrote later to his friend Winthrop Bell, with whom he had debated the rebuilding of Germany after the First World War. "They all hide something before themselves and even more so from others." But he did not have the heart, he noted, to inquire in many cases about activities and motives during the war.

To his surprise he continued to hear favorable reports of many who were *belastet*. Once he even heard a strong defense of a man who had been the president of a People's Court but was sympathetically looked upon by people whom Courant himself approved. He asked in his journal: *"Where then are the bad Nazis?"*

He did not have occasion to see Bieberbach; but in Berlin he met Hasse, who was living in the American zone and lecturing at the University of Berlin in the Russian zone. Courant noted only: "Met Hasse. Mixed feelings."

In Berlin Courant also saw his brother-in-law, Wilhelm Runge, who in 1916 had helped him install devices for earth telegraphy on the Western Front.

In 1945, in the primitive conditions of the devastated capital after the war, there had been a desperate need for matches, which were no longer being produced. As soon as power lines were in operation, Runge and a few other men from Telefunken had begun to manufacture a little transformer which could be plugged into the line with a switch on the secondary and a coil made out of a paper clip. When switched on, the glowing paper clip was sufficient to start a fire, which could then be used to boil potatoes. With this beginning Runge and his colleagues had set out to rebuild Telefunken.

Courant thought his brother-in-law's optimism was "more temperamental than rationally founded." Still, he could not deny an urge to help people like Runge, Rellich, Springer, Reidemeister, and others with such "constructive energy."

Before leaving Germany for Copenhagen and the Bohrs, Courant found time to see such old Göttingen friends as Klein's daughter, Putti Staiger, who because of her loyalty to her many Jewish friends had been removed by the Nazis from her position as the director of a gymnasium for girls in Hildesheim.

He also learned the tragic fate of his cousin Edith Stein. She had become a Catholic in 1922 and had later taken the vows of the barefoot Carmelite order. In 1942, with her older sister Rose, a lay member of the order, she had been snatched by the Gestapo from a convent in Holland— part of a general retaliation against the Dutch church for its criticism of Nazi anti-semitism in Holland. Both sisters had died in the gas chambers of Auschwitz.

(Since the death of Edith Stein [Sister Teresa Benedicta of the Cross], there has been a continuing movement to have her declared a saint. Although Courant was deeply affected by her death, the incongruity of a Catholic saint among the descendants of Salomon Courant tickled him; and in later years he was always to refer proudly whenever possible to "my cousin, the saint.")

Back home in August 1947, Courant and Natascha sorted out their impressions of their journey. As they had expected, they had not discovered in Germany anything in the actual development of computing machines comparable with the results attained or likely to be attained before long in the United States. It nevertheless seemed to them that German skill and manpower might be utilized to help in the American development. On the human side, they wrote to Weaver at the Rockefeller Foundation:

"In spite of many objections and misgivings, we feel strongly that saving science in Germany from complete disintegration is a necessity first because of human obligations to the minority of unimpeachable German scientists who have kept faith with scientific and moral values. . . . It is equally necessary because the world cannot afford the scientific potential in German territory to be wasted."

"Today it seems difficult to imagine how important it was that [Courant] came back to the University of Göttingen as soon as the circumstances permitted," Claus Müller of Aachen said years later after Courant's death. "I still clearly remember the tremendous effects of his first visit."

In America, after his return, Courant continued to face the same problems he had faced during the first post-war years. His group still had no stability in relation to the university, but as a result of support by the Office of Naval Research there was a large degree of financial flexibility. In addition, to encourage the development of centers of applied mathematics in the country, the Rockefeller Foundation had made both NYU and Brown a generous yearly grant for a period of five years. Combining this money with contractual and university sources, Courant was able to find places for a number of gifted young people when they were brought to his attention.

One of these was Harold N. Shapiro, a student of Artin's at Princeton. Receiving his degree in 1947—a year when jobs for Ph.D.'s in mathematics were especially numerous—Shapiro had elected to remain at Princeton another year. Then, when he went to look for a job in 1948, jobs were suddenly very scarce. It is Shapiro's guess that Artin, who regularly gave courses at NYU, told Courant about him.

"Anyway, I got a phone call from Courant, whom incidentally I had never met. A little voice at the other end of the line said, 'This is Courant,' and then there was complete silence. I recognized the name, but I didn't say anything. Then after what seemed an endless silence, he said, 'I understand you are looking for employment. Do you think you would like to come to NYU?'"

For Shapiro—who later spent a number of years in business—the interesting thing about the way in which Courant hired him was that it represented the kind of spirit in which he did things.

"Courant gambled. He understood human beings and life. He felt that if he took somebody who didn't work out, he'd find some way to get rid of him—which usually meant getting him a job somewhere else. If he couldn't get him a job—it was sort of a joke but it was also true—he would simply move him to the office farthest back in the building—and the Bible House which we occupied then was a huge building—and hope that he would go away. Courant was 70 per cent from the outside world, 30 per cent from the academic world. He brought to the academic world this sense of timing, gambling, change and so on. And that's what made the institute."

Shapiro feels that he had an advantage in getting along with Courant during his first years at NYU because he had friends who had had unhappy experiences with him and he was thus able to avoid their mistakes.

It often happened, he told me, that Courant would become very close to a young mathematician and would shower the youth with attention, praise, support.

"He was a kind of public relations manager, too. He had this wide circle of friends in mathematics. It was thrilling, you know. I remember once I had published some paper, and he came to me and said that Siegel had been his guest over the weekend and he had explained to Siegel what I had done. Siegel knew? He cared? He had read my paper? It was fantastic!"

A sort of father-son relationship would develop between Courant and the favored young man, the "son" apparently unable to do any wrong. Then there would come a point when, instead of waiting for Courant's altruism—"which, by the way, you could always depend on"—the youth would presume on the relationship by asking for something, or by asking for too much. Such a request would always provoke a negative reaction on Courant's part. "He wanted to be benevolent, but if *you* asked him for something, he would freeze up."

The result, according to Shapiro, was not merely the rejection of the request but also, often, total rejection of the young man who had made it. In Courant's files I have several times seen letters which reveal such incidents—a young man writing bitterly, "Since [such and such a specific date] you have changed completely toward me"—and the helpless question, *"What have I done wrong?"*

There were also often young women students over whose studies Courant would fuss constantly. One of these was Anneli Leopold, who is now the wife of Peter Lax.

"When I was very young, I almost felt about the Courants as if they were another set of parents," she told me. "They would invite me to stay at their home, and always Courant was interested in what I was doing, my studies and so on, a very fatherly kind of interest, also very stimulating to me."

After the war, when he had more financial flexibility, Courant sometimes added to the group attractive young women who had no special job qualifications. Somehow, though, he always managed to find useful tasks they could perform. They were referred to as Courant's "flames"—the term which had been applied in Göttingen to Hilbert's many romantic attachments. The true "flames" were, like Natascha Artin, intelligent and attractive women who were to continue to consider Courant their very good friend long after his intense interest in them had passed. They were always drawn into his mathematics enterprise and always became an integral part of the group. Jerome Berkowitz, who came to the Bible House in 1949 and observed the "flames" over a period of more than twenty years, told me that it seemed to him that Courant utilized his women friends to extend his own grasp of life. Friedrichs, who admits he never really paid too much attention to such matters, sees in the "flames" yet another example of Courant's "ability to be fascinated"—"Courant was fascinated by very rich men, he was fascinated by gifted young mathematicians, and he was fascinated by attractive and intelligent young women!"

"Has anyone spoken about Courant and women?" Louis Nirenberg asked me at the end of our conversation. "If you're not there and you don't see it, you just think, here is this old man and he's enamoured with this young girl, isn't that silly; but when you see what it does to him, it's not silly. It transforms him. You know how nice it is to be in love. Well, it is just as nice for him at an old age as it was thirty or forty years before. It is a constant thing. There is always somebody filling this role for him to a different degree—the degree varies from person to person. I don't find it silly. And I don't laugh at it. He just seems more alive at that time and more giving, not just to that person. It's touching and beautiful. Well, you have to see his face."

Discussing the many women friends of her husband, Nina said to me quite simply, "He needed them." For the most part—perhaps with the example of Mrs. Hilbert before her—she brushed them away as she did the moroseness and rudeness which he often showed at home. After his death she made it possible for the two young women who had been important to him in his last years to fly to the memorial service held in his honor at NYU.

A young woman who came to the Bible House at approximately the same time as Shapiro came was Lucille Gardner, now Wolff. She is the younger sister of Clifford Gardner, who was then a member of the group. As soon as Courant learned that Gardner had a sister who was a gifted violinist, he insisted that the young man bring her out to New Rochelle.

"So we went out to New Rochelle, and there we had the family chamber music, which was pretty astonishing for anyone," she told me. "Mama always knew what note it was and where we should be and so on. Papa, on the other hand, played really beautifully but was all wrong half the time. Imagine the young student arriving and there is the family. One person understands everything—not the mathematician. The mathematician doesn't know what beat he's on, he is just—in the most romantic way— playing away, whereas the musician in the family knows all about the beats but wrecks everything by coming in and saying, 'Papa! B!' Then he would come out of his fog and go, 'Ah!'—sort of a gasp, very humble. He was always wrong. He did not play at all intellectually. He simply had *no* idea!"

After meeting Lucille Gardner, Courant immediately set out to find some sort of employment for her, beginning with mechanical drawing and ending—when it was evident that she couldn't draw a straight line—with editorial work on the English translation of Courant-Hilbert. She just happened to be proficient in German. She told me she had admired the beauty of the Courant calculus, and she was stunned by the way in which the translation and revision of the classic of mathematical physics was being prepared.

"Nobody but Courant could have done that, putting together a book by farming it out to twenty different people. He let everybody put in their two cents' worth, and then somehow he pulled it all together. Only he didn't really pull—and yet somehow in the end it was all really together."

It was very much like other things she observed at the Bible House: "Papa—Courant—collected people and he collected ideas and he collected things and he managed to put them together. He ran everything in a way that you are brought up to think is the wrong way. The amazing thing was that the operation worked, and it worked because he was in the center of it, not letting go of any of the threads really, not really letting go. I wouldn't have believed it ahead of time, and it's hard to believe in retrospect!"

Lucille Gardner stayed at the Bible House until her marriage to a man much older than she, upon whom she had set her heart even before her meeting with Courant. While she was with the group, she did a lot of rewriting and editing, struggled to improve Courant's innumerable memorandums on the *reservoir of talent* which was *out there* waiting to be *tapped* by NYU; helped Natascha with the editing of a newly established house journal; burned secret documents.

"So it turned out I wasn't really a handicap; but if I had been, I wouldn't have been fired anyway."

After the second war, as after the first, Courant was much concerned with scientific publishing. He continued his collaboration with Interscience; and in some ways, Proskauer told me, Courant and he established the relationship in this country that Courant had had with Ferdinand Springer in Germany. But although they were "close and dear" friends, the relationship was not really the same; for they were of a different generation.

"Courant always drove a very hard bargain in his sweet way, but he was always very protective of Springer's rights, very alert to the slightest disadvantage to Springer," Proskauer went on. "After the war he worried a lot about the future of the Springer firm and even tried to bring Springer to this country to start a new publishing firm here. He thought if Springer did not survive, it would be a great loss to science. Also—of course," Proskauer smiled indulgently, "Courant always wanted to be a big wheeler-dealer in the business world."

After the war Courant had the idea that the reports which his people were regularly making on their work for the ONR could be printed in a journal instead of being mimeographed for limited circulation. The work could then be shared with the entire scientific community; there would be a better job for Natascha Artin, as the managing editor; New York University would have the prestige of publishing its own mathematical journal. All of this could be legitimately subsidized by the ONR. With 200 firm government subscriptions the first issue of NYU's *Communications on Applied Mathematics* appeared in January 1948. (Originally intended as a house journal, the *Communications* very shortly began to invite "outside" contributions and by its fifth issue had changed its name to *Communications on Pure and Applied Mathematics*.)

That same year Interscience also began to publish a series entitled *Monographs in Pure and Applied Mathematics*—an American counterpart of the famous Yellow Series, but a "non-yellow" series, the bindings always red and gray. (The first volume was the Courant-Friedrichs *Supersonic flow and shock waves*.) Ultimately, the series absorbed another series, the Interscience *Tracts in Pure and Applied Mathematics*, which Lipman Bers, then at NYU, was active in developing.

In the summer of 1948 Courant visited Germany for the second time since the war.

By then, the intense economic, political, military, ideological conflict between East and West which was known as the Cold War had resulted in a considerable change in the attitude of the western victors toward West Germany, which was now looked upon more as an ally than as a defeated enemy.

After the trips of 1947 and 1948, Courant visited Germany almost every summer until his death. He did everything he could to assist Rellich in rebuilding the Mathematics Institute, accepted membership again in the Göttingen Academy of Science, and became again an editor of the *Mathematische Annalen*, served informally as scientific adviser to Springer, participated actively in scientific and educational reconstruction, brought a number of German mathematicians as guests to NYU.

He never considered returning permanently to Germany, although some of his friends did. Siegel assumed his old chair in Göttingen. Artin became a professor again at Hamburg and resumed his old friendship with Hasse. Max Born, after his retirement from Edinburgh, spent his final years at Bad Pyrmont, a little spa not far from Göttingen. On the other hand, many of Courant's German-born colleagues, most notably Einstein, were never able to bring themselves to go to Germany again, even for a visit.

"There is no question but that after the war Courant was absolutely committed to the United States and to his enterprise at NYU," Friedrichs told me, "but he simply couldn't resist going back to Germany as soon as possible. In spite of what had happened to him, he wanted *to help*. He was very much criticized for that by many people. But such was his nature."

TWENTY-SIX

PEACE WAS NOT peaceful. The victors separated into two opposing camps. Berlin, divided into four sectors within the city but situated entirely in the Russian-controlled sector of Germany, became a pawn in the struggle between East and West. It was a time of the Truman Doctrine, the Marshall Plan, the Berlin Airlift, the North Atlantic Treaty Organization, the civil war in China, the retreat of the government of Nationalist China to Formosa, the inevitable explosion of an atomic bomb in the USSR.

As a result of the war and the ensuing "cold war," half a dozen centers of applied mathematics had sprung up in America. The subject had experienced a kind of boom. Other universities began to look acquisitively upon Courant's group. In 1949 there were offers to take over the operation from NYU and set it up as an institute for advanced study and research in the mathematical sciences on a new campus. To Courant such a development was not only flattering but also in line with his idea of centers of scientific research to be located in various regions of the country. He was particularly gratified by an offer from the University of Maryland because he had long advocated a development in or near the nation's capital. And still—". . . after much soul-searching, I must say that, at this moment, the proposed shortcut of simply transferring our present institute from New York University to the University of Maryland does not seem to us the most suitable way to implement your general plan," he wrote to Dr. H. C. Byrd, the president at Maryland. "It is not only natural inertia but, even more, loyalty to our mission in the New York region and to our institutional background which makes us hesitate simply to accept your generous and very tempting offer."

Yet, after fifteen years, the problem of "stabilizing" the group at NYU remained. The word began to crop up with increasing frequency in Courant's letters and memorandums in spite of the fact that his secretaries red-penciled it as often as they could. There were also increasing attacks from the outside. Max Shiffman, whose promising career was ultimately to be blighted by a long and serious illness, accepted a position at Stanford. Flanders finally decided to give up the academic life and join the scientific staff at the Argonne Laboratory. Several people went into industry, where they sometimes received twice what they had been paid at NYU.

The Graduate Center of Mathematics was still a poorly supported unit of the university, inadequately housed, and with just three professors whose primary appointments were in the graduate department. Basic salaries of these and of others were paid only in part by NYU. To a large extent even money for such academic necessities as fellowships, books and journals, desks and other equipment had to be obtained by Courant from outside sources or by juggling contract funds which in principle were not to be used to meet academic needs.

The mathematicians were beginning to feel like the stepchild of fairy tales who is sent out to work but whose wages are confiscated at the end of each day. The University of Maryland proposal to take over the applied mathematics group had been turned down in a couple of weeks. A similar proposal later in 1949 from the University of California could not be refused so quickly.

In the years that Courant had been at NYU, Griffith Evans had built up at Berkeley one of the top mathematics departments in the country. He was one of the few outstanding native-born American mathematicians who worked in the field of partial differential equations as well as one of the few who had encouraged Courant's efforts at NYU in a friendly fashion. Evans had recently retired and had been succeeded by Charles Morrey, a mathematician for whom the people at NYU also had tremendous respect.

Friedrichs still recalls vividly the first time he met Morrey.

"It was one of the New York meetings of the American Mathematical Society about 1938; and this inconspicuous-looking boy came up to me and said modestly that he wanted to tell me he had been working on partial differential equations and he had solved such and such a problem. I said that was very nice, and went on. Then—wait a minute!—I suddenly turned and went back to him and said, 'Would you tell me once more which problem it was you said you had solved?' *I couldn't believe it.* It was one of those problems many of us had worked on for years and years—I just couldn't believe it. Oh yes, Morrey is powerful!"

Thus, when Courant was hiking with Morrey and Hans Lewy in the Berkeley hills one afternoon in August 1949 and Morrey wondered aloud whether the NYU group would not like to transfer its operation to Berkeley, the idea was attractive to the New Yorkers.

Things began to move very rapidly. Originally—Morrey told me—he had been thinking about getting just Courant, Friedrichs and Stoker; but the proposal quickly ballooned into the transfer of nine full and part-time faculty members plus twenty student assistants and computers, one secretary and three typists from New York to Berkeley.

Even back in the reality of the Bible House, Courant could not close off the Berkeley proposal as quickly as he had the one from Maryland. The lure of Berkeley, "general and personal," as he described it to Morrey, was too great. He saw a possibility (which he later conceded was "perhaps naive") that financial support in Berkeley could be obtained directly from the State of California and his group would then no longer have to compete for funds with other university departments or to beg for support from wealthy citizens and private foundations. He visualized, he wrote to Morrey, "an institute for advanced study"—such as he was trying to establish at NYU—on the west coast as well as on the east.

Throughout the fall and winter he debated the move with himself, with Friedrichs and Stoker. He sought the advice of friends on the Board of Trustees, of Warren Weaver and the Rockefeller Foundation, of admirals and important government officials in Washington. He wandered around the Bible House, looking helpless. Everybody, down to the most recently hired clerk, was asked for an opinion.

"What we would wish," Courant explained to Morrey, "is to find in discussions with you a way in which [the] apparent conflict can be resolved so that we can help build up the institute in Berkeley without destroying what has been developed in New York."

By the end of November, although the original proposal of moving the entire group to the west coast had not been ruled out, Morrey was suggesting as an alternative that there could be a cooperating group on each coast with two principals in New York and the third principal and Hans Lewy in Berkeley.

"It was a typically Courantian idea and naturally it appealed to Courant," Friedrichs told me, "but Stoker and I were not in favor of it."

Although Courant told Morrey that he did not feel the Berkeley offer should be used to exact concessions from New York, he had already taken the opportunity to list for Chancellor Chase fourteen specific suggestions for improving the situation of the group at NYU. These included a personal request from Courant himself, approaching his sixty-second birthday:

"Orally I [have been] assured that I would not have to retire at the age of sixty-five. In view of the fact that this age limit does not apply at the places from which I have received recent offers, I should appreciate that such an assurance be given to me in writing."

"Courant would never leave a simple situation alone," Bromberg told me. "He always liked a maximum of unstructured developments. In a state of confusion, where others were most uneasy, he was most receptive to the significant aspects of the situation; and he always felt that much more was accomplished in such a state."

He had a "fantastic" sense of timing, in Bromberg's opinion, and he would use it to throw people off balance in even the simplest situation. As an example Bromberg recalled Courant's talking to someone in his office—"particularly a younger person who was probably in the dark about why he had been called in anyway." Someone else would stick his head into the office—the door was almost always open—and, seeing that Courant was occupied, apologize for the interruption and start to back away. "Come in, come in," Courant would mumble and, turning to the person with whom he had been talking, say, "We have no secrets, have we?" Just as the newcomer started to enter, Courant would murmur apologetically, "But perhaps—" with a look across his desk—"you would prefer?" As the visitor turned a second time to leave—"No, no, don't go—we have no secrets here, have we?" to the person opposite him, and then just as the visitor once more started to enter the office—"Unless you prefer?"

This sense of timing is very evident in Courant's correspondence about the proposed Berkeley transfer; and as the academic year 1949–50 progressed and the various alternatives were discussed and rediscussed, the group in the Bible House moved step by step to a schizoid frenzy.

In New York, as late as March 1950, secretaries were being hired on the condition that they would be willing to go to California in September; but in Berkeley the developing controversy over the loyalty oath required by the regents was beginning to tear that faculty apart. By May the negotiations for the transfer were faltering, and—as Friedrichs told me—"even before then, we had definitely decided we would not leave New York."

A question still debated by those involved is whether Courant *ever* had any real intention of transferring his group to Berkeley.

There are many who feel that he merely liked to play with the scheme, stir things up, pit Berkeley against New York to improve the situation there.

Morrey feels that the transfer of the group—in whole or in part—was seriously under consideration.

De Prima, who was at Caltech and thus not personally involved, says that Courant was honestly torn: "He called me several times and asked would I please come east. He just wanted somebody to talk to about Berkeley, the pros and cons."

Stoker, who had been most enthusiastically in favor of the move to Berkeley, told me that this was the only time he seriously considered leaving the group.

"I was so disgusted with the stingy ways of NYU that I wanted to go to Berkeley," he conceded, "although now I'm not sure that it would have been the best thing. Here Courant was his own boss. He went directly to the heads of the university and I did, too, afterwards when I became director. We could never have done that at Berkeley. And I don't think Courant really wanted to go. He didn't like the idea of leaving New York altogether. He felt this was the proper base for him."

"Courant really wanted to be at both places. That was what he always wanted," Friedrichs told me. "Even when you drove with him, you were aware of that quality in his nature. When he came to a fork in the road, he would quite literally *want to go both ways*. You could actually feel that!"

One of the attractions of Berkeley for Courant may well have been the powerful position it had in physics in 1950.

"It was one of our regrets that there was a poor physics department here at NYU," said Joe Keller, who came to the group after the war and is now the director of the Division of Wave Propagation and Applied Mathematics at the Courant Institute—the successor to the Division of Electromagnetic Research, developed by Morris Kline after the war.

In this connection Keller recalled his first trip to Europe with Courant around 1950 (in the course of which Courant arranged that Keller learn to ski by outfitting him, giving him some minimal instruction, and sending him down the slope). Although the primary purpose of the trip was to find out what was being done in Europe with explosives, it was also a talent search. In Göttingen, Courant met Bruno Zumino, a young Italian who was a student of Heisenberg's. A year or two later he brought Zumino to NYU, where he eventually became head of the physics department and instrumental in building it up.

Keller feels that if the NYU group had gone to Berkeley, as proposed, it might have had more impact on mathematics in the United States than it has had.

"The point I want to make is this. Our institute has been eminently successful in its field of analysis and applied mathematics, and it is undoubtedly the leading place in the world in those two disciplines. However, it has not had a corresponding impact on mathematics in the United States. Throughout the United States applied mathematics is given short shrift. The parts of analysis that *are* emphasized verge on the pure—what we call 'soft analysis' in contrast to 'hard analysis,' which is closer to the applications. If we had gone to Berkeley in 1950, we would immediately have become part of the leading mathematical establishment in the country and we would have had about a ten or fifteen year headstart on what we have been able to do here."

By summer 1950, however, the pros and cons of transferring to Berkeley were quite definitely a thing of the past for the group at NYU.

Courant was finally bringing to a conclusion his book on *Dirichlet's principle, conformal mapping, and minimal surfaces*. Almost all of his mathematical work had been one or another variant of Dirichlet's principle; yet he felt that many attractive questions were still open. He hoped and expected that his book would provide a stimulus for further research in the field.

The book is, according to Friedrichs, the most "Courantian" of all Courant's books in subject matter and the least "Courantian" in style and treatment.

"It is very sober, very concise mathematically. There are many technical details, delicate points, attractive geometrical situations and considerations. I would say that mathematically it is probably his best book. Yet of all his books, it has had the least impact."

"It is very difficult to say why it didn't register more," Courant himself said some ten years after its publication. "It was maybe not well written. Or not written with sufficient fanfare. As one gets older, of course, things that are close to one's heart go out of fashion. At least temporarily."

The same summer (1950) that Courant was completing the book on Dirichlet's principle, the first international mathematical congress since 1936 was held in Cambridge, Massachusetts. Originally scheduled for summer 1940, it had been the special project of Birkhoff, who had seen the first official congress on American soil as an event which would signify America's arrival on the international mathematical scene. With the declaration of war in 1939, the planned congress had had to be postponed until all the mathematicians of the world would again be able to attend. Birkhoff, who died in 1944, did not live to see that time.

In 1950 there were still many difficulties in bringing together the international mathematical community. In June of that year the United States sent air and sea forces to the support of the Republic of Korea. The USSR refused to permit any of its mathematicians to go to the United States. The American State Department looked askance at some mathematicians who wished to come. A number of Americans had strong reservations about the extent to which they would associate with mathematicians whose record in relation to the Nazis could be questioned.

As a result of the Berkeley offer, Courant had managed to obtain some money from NYU to invite some of the foreign mathematicians who came to the congress to stay over and give lectures. One of those he invited to give a few talks was Heisenberg. Another was Franz Rellich, who lectured at NYU for a semester.

Rellich was the first of a number of German scientists to be brought to NYU in the coming years. A more recent visitor has been Martin Kneser, a later director of the Mathematics Institute of Göttingen, the son of Hellmuth Kneser and the son-in-law of Hasse.

The invitations to German scientists were always the subject of considerable uneasiness in the faculty. Some people objected to anyone who had remained in Hitler's Germany being invited to NYU. Others made distinctions of different degrees. In the half a decade since the end of the war, Courant himself had found it increasingly difficult to draw a sharp line between "good" and "bad" Germans.

"Courant is amazingly tolerant of flaws in character—in a way that, you know, I couldn't be," Anneli Lax told me. "He seems to look at the white and the grey and the black of a person and then put weights according to his own values. He can excuse things in people which most of us could not excuse. At the same time, somebody like Flanders, who was above reproach as far as character is concerned, is an object of great respect for Courant."

"I personally don't have the strength of character [of someone like Einstein] that I can hate people or a country so absolutely without reservation that it influences my whole existence," Courant explained to me during the last year of his life. "One cannot forget all the suffering, of course; but I was always in favor of being positive toward Germany, reestablishing contact. I felt very bad about Bieberbach once, but now it's been so long—so much time has elapsed—and I can always see that there were mitigating circumstances with all these people. I never had any contact with Bieberbach again, but if I would meet him here now I would be friendly with him."

Following the 1950 congress, Laurent Schwartz also gave a series of talks at NYU. Schwartz's theory of distributions was considered a conspicuous example of the trend toward generality and abstraction in mathematics which Courant had opposed throughout his career and from which he always tried to "shield" his students.

"There was a certain amount of provincialism at NYU, which was somewhat Göttingen-like," Peter Lax explained to me. "I guess Friedrichs told you how von Neumann was considered a flash in the pan with his operator theory there. Too abstract and all that. Well, we felt the same way about Schwartz's theory of distributions. It is one of those theories—there's no depth in it, but it's enormously useful—and it's different from the Hilbert-space approach in which Friedrichs pioneered—so we resisted it a little bit."

In spite of this resistance, Schwartz was invited to talk at NYU. A few years later, using Schwartz's theory, Lax found a way of proving some things about the solutions of differential equations and, as he said, "that certainly changed my mind and I think it changed Friedrichs's mind and Courant's mind too. In particular—I am jumping ahead now—the last big scientific effort Courant made was to write an appendix on distributions for the English edition of Courant-Hilbert."

Courant often spoke out very strongly against the trend toward purism, generality, and abstraction in mathematics. Once, when I saw a reference in a speech of his to "the blasphemous nonsense of 'mathematics for its own sake,' " I asked him if this didn't represent a considerable difference of opinion between him and Hilbert—who on occasion admiringly quoted Jacobi's rebuke to Fourier: "A philosopher like Fourier should know that the sole aim of science is the glory of the human spirit."

"With Hilbert there is really no contradiction," Courant replied promptly, "because Hilbert didn't live to see this overemphasis on abstraction and the self-emulation and self-adulation that some of these abstractionists show."

"We at NYU recognized rather tardily the achievements of the leading members of 'Bourbaki,' " Friedrichs explained in this connection. "We really objected only to the trivialities of those people whom Stoker calls *les petits Bourbakis.*"

Lax was one of the last students to take his Ph.D. with Courant. He found him "a very original guy" whose way of expression matched his way of thinking.

"The way he wrote was like nothing anybody else wrote. He hated that style of stating a theorem which goes, *Let M be a manifold, X a differential structure, Y a vector field, and so on.* He would really want to describe first what the problem was about and how one goes about attacking it. In fact, perhaps Courant went rather more in the other direction of not having enough theorems. But—as he said—'Most people have too many.' He was pretty old by the time I started working with him, but by far the most interesting work we did together was when he was sixty-eight. We generalized Huygens's principle."

I asked Lax how he and Courant had worked together.

"It came about this way. In Courant-Hilbert the propagation of discontinuities was worked out for second-order hyperbolic equations. Courant thought it should be worked out for general systems. I thought that could be easily done, and I did it—it went along the same lines as in the second order case. About the same time Courant had written a paper with Anneli on solving the initial-value problem. It was something that worked only for equations with constant coefficients, but I observed that putting together what Courant and I had done for the propagation of discontinuities with the technique described in that paper of his and Anneli's, one could get, not an exact formula for the solution of general hyperbolic equations with variable coefficients, but a formula that gives correctly the propagation of singularities. And it worked out very simply."

"But it sounds as if you did most of the work."

"No. Because, first of all, it was Courant who said the propagation of discontinuities should be worked out for the general equations and, secondly, it was he who with Anneli worked out the formula for the solution of the initial-value problem—which then could be put together."

"So you don't think it was an instance of Courant's having the first night, as it were, with one of his students?"

"No, definitely not," Lax laughed. "I would go out to New Rochelle to talk to him—in fact, on this particular problem I saved several pieces of correspondence, Courant's scribbly handwriting—it was a very natural and easy collaboration."

"Courant has always been completely generous in providing ideas for what you should do," said Jerome Berkowitz, who was a student a little later than Lax. "He was quite uninhibited—which most mathematicians are not, by the way—about telling someone else, not just a student, what he ought to do. Of course young people were especially susceptible to this; and he had really good ideas about what they should do—also their talents—and steered them into fields which certainly they would not have gone into otherwise.

"In his courses he would always give you the big picture. He had very good instincts about what could be done and how you could get from here to there; but when it came to actually doing it, his handwriting on the board would get worse and worse and every symbol would look like the same Greek letter, and things would be erased, and he would say, 'And there it is!' But he would make the idea of what you had to do to achieve it very clear, and you would then do it."

That September of 1950, when Rellich and Schwartz were delivering lectures at NYU, enrollments at all American colleges and universities were sharply down. This was partly due to the number of young men being drawn off by the armed forces, but also to the fact that the 18-year-olds of that year had been born in 1932, the worst year of the Depression, and were as a consequence relatively few in number. At NYU alone the drop in enrollment represented a loss of $1,600,000 in tuition fees. Nevertheless, as a result of the possibility that all or part of the Applied Mathematics Group might go to Berkeley, the general situation of Courant's people had improved.

"Under the prodding from various agencies"—in Courant's words— the university had become "more actively concerned with the material welfare of our enterprise." There had been a substantial increase in the budget and the university had assumed a larger share of basic salaries. But, as Courant wrote to Morrey in one of their last exchanges, the problem of *stabilizing* the group had not really been solved "and probably is not solvable in the structure of NYU."

In the fall of 1950 Courant began to ponder once again his proposal for some sort of national institute of the sciences. The need seemed even more urgent now than it had before the war. The universities of Europe were no longer what they had been. The American universities had "a major responsibility as guardians of the intellectual values of Western Civilization." The principles upon which Flexner had founded the Institute for Advanced Study should prevail, but—rather than a single and exclusive scholarly retreat, reaching only a small group of students—there should be "facilities for freely advanced learning . . . created at many places, [unhampered] by the multitude of educational services which, in a democratic society, have to be rendered on a non-selective mass scale."

During the Korean War a new series of lengthy memorandums began to go out from Courant's desk—some paragraphs carried over word for word from the original memorandum drawn up in the winter of 1939–40. The aging Flexner, eighty-five and loosely attached to the Carnegie Foundation in an advisory capacity, was still interested in Courant's ideas; and Courant tried hard to interest Robert Oppenheimer, the new director of the institute in Princeton.

"I could envisage great benefits for higher education if the Institute for Advanced Study would take the leadership in such an effort," he wrote to Oppenheimer.

But Oppenheimer did not respond.

In the middle of December 1950, President Truman declared a state of national emergency and outlined plans for placing the country again on a war footing.

Courant scurried back and forth between New York and Washington. He saw to it that there were always some of "his people" working for the ONR. He was constantly alert for connections between students at NYU and names in the daily headlines. His circle of acquaintances in government, military, and business—already large—continued to widen. Friedrichs and Stoker assumed more and more duties and responsibilities for the group. Stoker's wife, Nancy, became Courant's administrative assistant, handling with efficiency many of the multifarious problems of the growing operation. And yet—as people in the group frequently observed to me—even when Courant wasn't actually at the Bible House, he seemed to be there.

It was just at this point that even the success of Courant's efforts as an advocate of applied mathematics in the United States began to work against him. For a number of years NYU's Graduate Center for Mathematics had been receiving an annual grant from the Rockefeller Foundation. At the end of 1950, a five-year grant of $12,000 a year having expired, Courant, Friedrichs and Stoker wrote a joint letter to Weaver asking that it be renewed and, if possible, augmented.

Weaver responded regretfully.

"Ten years ago the situation in this country with respect to applied mathematics was very different from what it is now." In 1940 there had been only two places which had qualified for help from the Rockefeller Foundation to encourage applied mathematics—New York University and Brown University—and it had therefore made an exception to its program by giving some modest help to both institutions. Since then the general attitude in America toward applied mathematics had changed. There were now several effective centers for training in the field. "You and your colleagues can take a special pride in that statement, for you have without any question played a large role in making that statement true." But the period of pioneering help—as an exception to the program of the Rockefeller Foundation—was over.

At the same time that this important source of funds was being cut off, the financial position of New York University was becoming increasingly precarious. The war in Korea continued to drain off enrollment. Chase had retired. No successor had been chosen. There had been virtually no implementation by the university of the recommendation of Roberts's committee for the establishment of a mathematics institute at NYU. Yet in spite of the many disappointments, as Courant reminded himself, "moral backing" by a number of distinguished people, among them Augustus B. Kinzel, an outstanding engineer who was the vice president of Union Carbide, had persisted. In 1951, with the encouragement of Kinzel and others, Courant came to the conclusion that what he had been trying to achieve for so many years at NYU—an advanced institute of mathematical research and training—simply could not be achieved within the framework of that university. Still once again, in the spring of 1951, he brought up his idea for a national institute.

But he couldn't convince the people he wanted to convince.

"If one thinks of [your proposal] concretely in the terms of New York University and Columbia, for example, just specifically why would this combined institute offer something which is not realizable within the natural framework of the separate institutions?" Warren Weaver demanded. It was certainly true that what Courant had in mind had not been realizable "in its best form" at New York University. "But again would there not be those who would say that this leads to a criticism of New York University rather than to a proposal to start something new?"

For that, it seemed, there was no answer.

TWENTY-SEVEN

I N 1952, with the appointment of Henry Heald as chancellor of New York University, Courant—"an incurable optimist," as he admitted—returned again to the idea that perhaps an institute of science could be achieved within the framework of NYU. A memorandum of April 15 mentions "an encouraging discussion with Dr. Heald" and the fact that it now seemed "proper to suggest" certain initial steps of implementation and development.

Then, just at this time, Courant heard a significant piece of news. The Atomic Energy Commission was planning to expand its utilization of the recently developed high-speed electronic computers and was planning to install a machine in each of its laboratories. It would then locate the single computer that it had previously bought—the UNIVAC 4—in some institution in the eastern part of the United States where scientists, removed from the pressure of the laboratories, could develop more efficient methods for solving numerical problems involved in the work of the AEC.

During the past few summers Courant had frequently visited Peter Lax when that young man was a consultant at Los Alamos. He had found the big machines fascinating, although he had not been able to go as far as many mathematicians in their enthusiasm for what computers could do. He sometimes referred to them drily as "the emperor's new clothes."

"Maybe because the evidence that was offered for computers then was bad," Lax said to me. "I think there were a few instances where things were offered—oh, such and such a calculation was done on a computer which couldn't have been done otherwise—and then it turned out—I remember one such instance—that Isaacson could do it with just a little theory and a few desk machines. However, the conclusion to draw is that, well, that was a bad example; there are other good examples.

"Computers can do marvelous things! Computing really gives new content to scientific theories and completely alters the nature of applied mathematics. Mathematicians like *general* things, and now for the first time with high-speed electronic computers we have a chance to work on general things in the applied field."

With the news that a generously-budgeted computer was going to be available, whatever reservations Courant had about the big machines evaporated. But before he made any move to try to get the UNIVAC 4 for his group, he consulted Friedrichs and Stoker.

"Courant would never have begun an enterprise if Friedrichs and Stoker disagreed with him as to the value of it," I was told by Isaacson, who had continued to run the NYU "computing department" in the years since the war. "He would try in his own inimitable fashion, indirectly, to get them to see the light so that they would come around to his point of view; but he would never in a direct way oppose a position which they had both taken. In this case 'education' wasn't necessary. They immediately saw that it would be a very good idea."

Thus began what is still remembered by Isaacson and others as "the battle for the computer"—the turning point in the history of what was to become the Institute of Mathematical Sciences at New York University.

Besides its value as a tool of applied mathematics, there was another important reason for trying to get the UNIVAC 4 at NYU. Because of the high initial investment in the machine and its placement, the program would of necessity be a continuing source of work and support which could possibly result in the long sought stabilization of the applied mathematics group. Yet, according to Isaacson, the farsightedness of Courant, Friedrichs and Stoker was somewhat surprising; for the project as originally envisaged by the AEC was a mere service activity. A minimal staff would run it, develop the necessary programs, and maintain the machine.

During the summer of 1952, the Atomic Energy Commission scheduled a meeting at which an AEC representative would present the details of the project to the interested institutions eligible to submit proposals for the use of the computer. At the time Courant was in Europe. Stoker, who was lecturing that summer at Stanford, flew to New York. He, Isaacson, and a member of the university administration represented NYU at the meeting in the Empire State Building.

The representatives of the various institutions were seated around an oblong table. The chairman took the 12 o'clock position. The NYU contingent was assigned the 1 o'clock position at his left. After the AEC representative had explained the nature of the project, the chairman went around the table, counterclockwise, and asked a representative of each institution to comment in turn. To the amazement of Stoker and Isaacson, the response of those who spoke was, without exception, negative.

"They said they couldn't possibly think of making a proposal unless they had the use of the machine at least fifty per cent of the time for their own purposes—at the AEC's expense of course."

When a break was called for lunch, only NYU had not yet been heard from. At lunch Stoker and Isaacson freely said that it was ridiculous for the others to feel the way they did. Although the immediate objective was to run a computing center for the use of the AEC laboratories only, it was not going to persist that way. After lunch, when NYU was finally called upon, Stoker expressed his opinion that for people interested in applied mathematics having the computer would be a great opportunity.

"We then went our merry ways," Isaacson recalled, "and, lo and behold, in spite of the fact that all the other people had objected to the terms, every one of the institutions which had been represented at the meeting submitted a proposal for the machine essentially on the terms the AEC had laid out."

Since the budget for the project had to be based on the minimal staff that Remington Rand had suggested, all the institutions proposed roughly the same figures. The people from NYU were rather confident their proposal would be the one approved because of their extensive experience in applied mathematics. A month or so later, however, after Courant had returned from Europe, they learned indirectly that the AEC was planning to locate the computer at another institution—oddly enough, one which had not been represented at the meeting at the Empire State Building.

Everybody was angered and deeply disappointed, but what was there to do about it?

Stoker simply refused to give up.

"I think Stoker deserves a great deal of credit for that," Friedrichs said to me. "I remember that he told Courant, 'You can't let them do that to us!' He simply *made* Courant go back and fight to change the AEC's mind, because—as we were told—the contract had not yet been officially awarded."

"Courant was a tremendous fighter," Isaacson contributed. "He got all the scientists and the experts in Washington that knew about our work to get in touch with the administrators of the Atomic Energy Commission —scientists in scientific jobs who as government employees could speak on a par with other government employees. And then there were other people he brought in. He was absolutely inexhaustible. He was thirty years older than I, and I was worn out after an hour spent with him. But he would go on at great length, drafting letters, calling people, planning the moves that would be necessary to overturn the decision—which still had not been announced officially."

Isaacson is not sure which effort of Courant's was ultimately decisive: "Apparently the combination of all of them convinced the research division of the AEC—and in the end they awarded the contract to us!"

The AEC insisted that the building in which the UNIVAC was to be housed must be fireproof. The Bible House did not meet this last requirement; but Chancellor Heald had assured Courant that if he got the computer, a suitable building would be acquired for it.

Suddenly, after nearly twenty years of constant effort—of letters and memorandums, schemes and proposals, hopes and fears, disappointments and small successes—an institute of the mathematical sciences with all the multifarious activities to be housed under one roof, as in the old days in Göttingen, appeared to be coming at last into existence at NYU.

Another series of memorandums began to flow from Courant's office to that of Heald, but these were now concise and specific:

"The activities of (1) the present Institute for Mathematics and Mechanics (about 55 people), (2) Professor Kline's electromagnetic research group (about 28 people), (3) the AEC computing facility (initially probably about 25 people . . .), and (4) possibly in the near future a group in statistics and probability ought to be fitted into a single framework to be called 'Institute of Mathematical Sciences.' "

On the morning of January 18, 1953—ten days after Courant's sixty-fifth birthday—the *New York Times* carried the headlined story: NYU WILL EXPAND MATHEMATICS UNIT.

Although the newspaper announcement mentioned "a Division of Computing Services" as a part of the new "Institute of Mathematical Sciences," there was no mention of the AEC's UNIVAC.

This was for a very good reason.

With Chancellor Heald's assurance, Courant had had no qualms about promising that NYU would be able to provide the type of building required to house the computer; and it had been so stipulated in the contract with the AEC, Isaacson told me.

"Then, just a day or two before or after—I forgot which—when the representative of the AEC came to meet Heald and officially inform us that we had been awarded the computer, the building deal fell through. The owner, consulting his lawyer about the sale, had discovered that it would pose certain tax problems for him."

Heald moved quickly. In a very short time he had managed to sign a contract for another, larger building at 25 Waverly Place. On January 25, 1953, the *Times* was able to announce: NYU BUYS BUILDING.

The new building had housed a number of now bankrupt hat factories. Some tenants still occupied the upper floors under lease, but the mathematicians were able to take over the basement and the first and second floors immediately. They began to make the necessary preparations to install the UNIVAC.

Remington Rand had sent a template for the concrete platform which would have to be constructed for the machine. This indicated openings for various wiring conduits and air conditioning ducts. The university architect drew up the plans. The platform was poured with the holes in position as indicated. The frame for the machine was delivered. Then it was discovered that the architect had turned the template upside down and his blueprint had been the mirror image of the correct placement of the necessary openings.

An emergency meeting was called. The engineers from Remington Rand came up to New York from Philadelphia. To everybody's great relief, they concluded after inspection that with a few modifications the existing platform could be used.

"But there was still some consternation," Isaacson went on. "At that time the machine was the sole computing facility of all the AEC laboratories. It had to be kept going. So for four months we had to operate it in the factory in Pennsylvania where it was still located. I commuted quite regularly to check on the progress of our engineering and technical people and on the way in which problems were being done. The machine worked continuously 24 hours a day and every day of the week. The delay with the flooring could have been a catastrophe if the repairs had taken longer. As it was, we were able to work straight through. It was a *very* exciting time," he concluded, a little wistfully.

Bromberg, who had left the group at the end of the war to go into industry, was asked to take over the management of the new facility. He had obtained his degree with Stoker while still working at the Reeves Instrument Company, a firm that manufactured analog computers; and he had worked with numerical methods in his own research. He also had experience and interest in administration. He felt, he told me, that "it was like the family calling you home," and so he returned.

Initially the computer was a classified project under armed guard and with access only to a minimum number of people who had been cleared by the FBI. Stoker was the first senior faculty member to use the machine. Almost immediately he and Isaacson collaborated on work involving water waves and meteorology which ultimately led to Stoker's work on flood control.

Courant was always particularly proud of this work by Stoker—"a very fruitful combination of down-to-earth and purely mathematical components"—because, among other things, it perfectly illustrated his belief that although science cannot be *directed*, it can be given *opportunity*. For a number of years some members of the NYU group had been concentrating on the theory of hyperbolic partial differential equations, which describe wave propagation phenomena. Other members had directed their attention to numerical methods based on finite-difference schemes. Both of these activities were to a degree outgrowths of the work done in Göttingen by Courant, Friedrichs and Lewy. When in the 1950's Stoker was asked by the U.S. Corps of Engineers to devise methods of predicting floods, he had conveniently at hand this body of theory as well as a high-speed computing facility.

During 1953–56 Stoker took the record of one of the largest floods of the Ohio River to verify whether the flows in the river could be calculated theoretically to the accuracy desired. He found that this calculation was possible for flows in the 375 miles between Wheeling and Cincinnati and for periods up to three weeks. The calculation stretched the capacity of the UNIVAC 4 to the limit, but required only 6½ hours of computer time. The results established that utilizing the theoretical method was not only feasible, but also much more economical than what might be called the practical method of building a large-scale model (costing as much as the UNIVAC 4) for each river or reservoir to be studied—a method used in the past.

Yet, in a way, Stoker told me, he always regretted the acquisition of the UNIVAC.

"I saw it was impossible not to try to get it because it's such an instrument for applied mathematics, but I also saw that it would mean an enormous expansion and a kind of destruction of the place the way I liked it to be. From that time on, it lost its coherence. Friedrichs and I practically never worked together on a long problem, because the senior people all had duties and so did a lot of other people. But I saw it was inevitable, so I tried to help it along."

By 1958, when Stoker succeeded Courant as director, the AEC had replaced the UNIVAC 4 with an IBM 704 computer; and the computing center had taken over four floors of a nearby building at 4 Washington

Place. As Stoker and Isaacson had predicted, the machine was generally available to the institute.

Although NYU was by that time one of the five laboratories carrying on the major portion of AEC research, its participation remained unique in respect to the role played by students and by more than token representation of faculty. In addition, more mathematicians at NYU were engaged in the theoretical aspects of Project Sherwood (the AEC's continuing effort to harness nuclear energy for peaceful purposes) than at all the other four centers combined. While the other centers were charged in general with such immediate objectives as engineering design and development, the assignment of the NYU group was to help the AEC understand in detail what was happening inside the devices, insofar as mathematical science could do so.

Neither Courant nor Friedrichs ever used the computer in connection with their mathematical work; however, the ideas of the Courant-Friedrichs-Lewy paper on numerical solutions of differential equations became basic as the use of the high-speed machines increased. In 1967, some twenty years after the wartime development of the computer and almost forty years after the publication of the original three-man paper, IBM devoted an entire issue of its journal to the impact of "C-F-L" on modern investigations in the numerical analysis of partial differential equations—"an outstanding instance of research undertaken for purely theoretical purposes turning out to be of immense practical importance."

After the installation of the first computer in 1953, the whole mathematical development at NYU began to move with a speed which "quite overwhelmed" Courant. Leaving in June 1953 to attend the 1000th anniversary of the city of Göttingen, he wrote to Harry Woodburn Chase:

"I certainly have not forgotten the sympathy you have shown to my interests from my very beginning at the university. If it hadn't been for your intervention at critical points, the Institute would never have reached its present state of development."

Courant's visit to Göttingen in 1953 was almost as significant for him as his first post-war visit. As a part of the anniversary celebration, Born, Franck, and he were to be made "honorary citizens" of the town. The decision to accept this status had not been automatic for any of the three men.

In 1933 they had all reacted to the personal repercussions of Nazism in different ways. Twenty years later, each arrived at the same decision as the other two. In a ceremony in the town hall, which still bore emblazoned on the wall of its cellar the ancient motto, *Extra Goettingen non vita est*— "Away from Göttingen there is no life"—they accepted the impressive scrolls which announced that they were *Ehrenbürger* of Göttingen.

When Courant returned to NYU that fall, preparations were being made for the move to the new quarters. Shirley Twersky, who had been working as a secretary, told me that she saw this as a good opportunity to leave and join her husband, a former member of the group, who was already on the west coast in a new position.

"I was supposed to stay around just another month or so and train another girl to take my place, because Courant required very special handling."

"What do you mean by 'very special handling'?"

"Oh, he liked to play little games. Little puppet-type games. They were used to set up a situation and see what the reaction would be. For example, when I did finally convince him that I was going to leave, he came in one morning, just ambled in, and said, 'Well, it's too bad that you're going to leave just when everybody no longer dislikes you.' Natascha, who was there, got very upset and said emphatically that she had never disliked me. But I just laughed. He wasn't really being malicious, but he was an observer of people. He liked to see what they would do."

It should be said, however, that there were others with whom I talked who felt that Courant *was* sometimes malicious in his human experiments —that when he saw two people who were close to each other, for instance, he would frequently try to sow a seed of dissension between them. He himself shrugged and said that if he could break up a relationship, it could not be a good relationship for those involved.

Those who were closest to Courant freely concede his faults.

"In his case one is quite aware, you know, that he's not perfect," Anneli Lax said to me. "He does some things that are really quite objectionable, and yet you like the man. Perhaps it is because in spite of the somewhat devious ways the kinds of things he really stands for and accomplishes are O.K. I mean, they are the things you would want to do too, although you would do them straightforwardly. I am not a person who believes that any means is justified by the end, not at all, but he seems to get away with it. Somehow."

". . . a never failing loyalty bound Courant's associates together," Neugebauer said after half a century of friendship. "He inspired an unshakable confidence in his profound desire to do what was right and what made sense under the given circumstances. His ability to create a feeling of mutual confidence in those who knew him intimately lies at the foundation of his success and influence."

The move to the former hat factory in the fall of 1953 represented for Courant the culmination of his almost twenty years of effort to bring his experience in Göttingen to bear upon the situation of science in America. Although he had earlier applied the name "institute" to his group in the Bible House, and before that to the group in the Judson Dormitory, this was the first time that an institute in the sense of the one he had had to leave behind in Göttingen would come into existence at New York University. It was also an important development in American mathematics. A few years later (1958) *Fortune*, featuring a pair of extensive articles on mathematics in this country, ran Courant's picture and that of Oswald Veblen of the Institute for Advanced Study, side by side, as "the founders of two great mathematical centers." New York University's Institute of Mathematical Sciences was described as "the national capital of applied mathematical analysis."

Curiously enough, to the people in the group the new "institute" did not seem like "such a big thing." They were going to move gradually into a larger building. All of the mathematical activities would eventually be brought together. There would be more money, a new name. But to Courant —it is clear from his correspondence and memorandums—the move was comparable to that made long ago to the new building on Bunsenstrasse. He postponed the dedication for more than a year so that Niels Bohr could deliver the principal address.

Since Harald Bohr's death in 1951, Niels Bohr "had entrusted himself a little bit more," as Courant put it, to his brother's old friend. On Courant's side there was no question but that he considered Niels Bohr, not only the most outstanding scientist, but also the most outstanding human being he had ever met.

"There is nobody in my opinion who can be rated as highly as Bohr," he told me. "Hilbert was very original, and Bohr had great respect for him; but Hilbert was a very different kind of personality. I would not even rate Einstein as highly as Bohr. And of course in Göttingen I met gradually very many outstanding people in different fields. I might talk of Poincaré —but nobody—Bohr was unique. He was on a different plane. It was not only breadth. It was his penetration of nature. He understood what went on in physical phenomena. He understood it the way birds understand the singing of other birds."

There was in any case no more appropriate choice for the speaker at the dedication than Bohr. He had always been generous in backing Courant's American efforts with his own great international prestige, and he and his brother had made the original suggestion that Courant approach the Rockefeller Foundation for assistance in erecting a mathematics institute in Göttingen.

During the first year of the existence of NYU's Institute of Mathematical Sciences, the American scientific community was aroused and divided by what came to be known as the Oppenheimer case.

In April 1954, when headlines announced the withdrawal of Oppenheimer's security clearance by the Atomic Energy Commission, Courant wrote immediately to the physicist and offered "to join the ranks of your friends who rally in defense of your character." But he did not give up his friendship with Admiral Lewis L. Strauss, the chairman of the AEC, as many of Oppenheimer's supporters felt he should. He wrote to Niels Bohr: "[The case of Robert O.] has so many sides that I personally cannot accept the prevailing simplified version and attitude."

"In general, one can say that multisidedness was an essential feature of Courant's *Weltanschauung*," Friedrichs said to me. "He hated one-sided statements. He was suspicious of them. He was suspicious of all final and sharply given statements and formulations."

After the Gray Board upheld the AEC against Oppenheimer, Courant deplored the bitterness which the decision engendered among scientists. He agreed with Strauss that the scientist as such had no more claim to public influence than any other citizen, but he felt that the government could not afford to disregard "the necessary psychological conditions of the scientific mind." Something should be done to reassure scientists of the government's fairness and reasonableness in handling security problems. He wrote a long letter to Strauss setting forth his suggestions for action by the AEC to relieve the tension. He also felt, as he wrote to Niels Bohr, that Oppenheimer should modify his attitude in the affair: "It would be much healthier if he understood that, while there is practical unanimity about the outcome of the formal procedure, there is also much criticism of him in many respects. By showing conciliatory humility, he could really become a great man and make a great contribution."

There was also a "security problem" closer to home.

Donald Flanders, who had finally given up his professorship at NYU for a job at the Argonne Laboratories, had become the subject of an investigation to determine whether, because of his friendship with Alger Hiss, he was a security risk. Asked by Flanders if he would furnish an affidavit of his loyalty, Courant promptly agreed. Shortly afterwards, an FBI agent came to Courant's office to question him further about Flanders, who persisted in visiting Hiss while he was in prison. Eyes sparkling behind his glasses, Courant looked across his desk at the agent sitting in the chair opposite him.

"Ja, ja, Alger Hiss," he mumbled. "That chair," he said, not quite inaudibly, "that chair you are sitting in, when he came to my office, ja, Alger Hiss used to sit in that chair."

The FBI agent quickly came to his feet and, as Courant was subsequently to tell the story, conducted the rest of the interview standing.

When Flanders's brother Ralph, by that time a senator from Vermont, demanded the censure of Joseph R. McCarthy by his colleagues in the senate, Donald Flanders was again investigated. The month before the dedication of the institute, Walter Winchell announced in his Sunday evening broadcast that "only last Thursday" Senator Flanders's brother Donald had said that he had no intention of going back on his friendship with Alger Hiss: "Significantly, ladies and gentlemen, Alger Hiss, in prison for lying about his Red connections, will be out in about six weeks and he can count on a very, very dear, dear friend working on America's most secret weapon—the atomic bomb."

A few years later Flanders was to take his own life; but he had lived to see the dedication of the Institute of Mathematical Sciences at New York University, for which he had in a sense laid the cornerstone that long ago day in 1933 when he had approached Oswald Veblen for advice about improving the mathematical situation at NYU.

The dedication ceremony for the new institute took place in the auditorium of Vanderbilt Hall on November 29, 1954, almost to the day a quarter of a century after the dedication of the institute in Göttingen. After appropriate remarks by various officials of the university, there was music by the Moyse trio. Then came the director of the new institute. For once, he did not repeat the message that Klein had distilled from the example of the École Polytechnique and the broad universal works of Gauss, Dirichlet and Riemann—the message to which Hilbert had given living embodiment. Instead he limited himself to a few remarks about his long relationship with the Bohr brothers and the role which Abraham Flexner had played "[in stimulating] the hope that at least in the field of Mathematical Sciences something parallel to the Institute in Göttingen could be developed in New York City."

Then Niels Bohr, speaking as only he could, even more inaudibly than the director of the new institute, referring gratefully to Courant as his "other brother" and "the man who opened up mathematics to physicists," announced his subject—"to indicate," as he said, "the guidance which the mathematical sciences through the ages have offered as regards our orientation in that nature of which we ourselves are part, and especially to stress the fundamental role which mathematical abstractions and the art of calculation have played in the development of the physical sciences."

Among the men and women in the audience there were only a few who remembered the similarly cold but bright winter day when the Göttingen institute had been dedicated—the way in which the spirit of Klein had seemed to be a living presence at the ceremonies—the frail old Hilbert saying delightedly in his sharp Königsberg accent, "There will never be another institute like this! For to have another such institute, there would have to be another Courant—and there can never be another Courant!"

Just when Courant's dream of an institute of mathematical sciences in New York became a reality, his dream of Göttingen's returning to its old position as a great mathematics center under the leadership of Franz Rellich was coming to an end. The year following the dedication of the new institute, Rellich, forty-nine years old, died of a brain tumor.

"Everywhere that memories of old mathematical Göttingen are still alive, Rellich's death has caused great distress and sorrow," Courant wrote sadly. "Centers of science are fragile organisms, sensitive to the loss of the personalities who promote and unify them. When good fortune puts a man like Rellich in the right place, such centers can blossom for a short time, as we have seen with admiration in Göttingen. . . . Throughout the mathematical world it is hoped that in spite of Rellich's death the Göttingen tradition will live on."

TWENTY-EIGHT

COURANT served as director of the Institute of Mathematical Sciences at New York University from 1953 to 1958—the same number of years that he had been permitted to serve as director of the newly built institute in Göttingen.

For his seventieth birthday on January 8, 1958, his friends established a fund to endow a series of Courant Lectures to be delivered every two years. Friedrichs was chosen to present the gift.

"One may think that one of the roles mathematics plays in other sciences is that of providing law and order, rational organization, and logical consistency; but that would not correspond to Courant's ideas," Friedrichs said at that time. "In fact, within mathematics proper Courant has always fought against overemphasis of the rational, logical, legalistic aspects of this science and emphasized the inventive and constructive, esthetic and even playful on the one hand, and on the other hand those pertaining to reality. How mathematics can retain these qualities when it invades other sciences is an interesting and somewhat puzzling question. Here we hope our gift will help."

There were some other honors in the year of Courant's retirement. New York University awarded him an honorary degree. The Navy presented him with its Distinguished Public Service Award. The Federal Republic of Germany decorated him with the Knight-Commander's Cross and the Star of the Order of Merit—an honor that surpasses even the Iron Cross, First Class.

In the fall of 1958, Courant's position as director of the Institute of Mathematical Sciences passed quite naturally to Stoker, who immediately began to put into effect his own long held ideas about the way in which the financial and administrative position of the institute should be set up in relation to the university.

Intellectually, Courant believed that when one retired, he should step down completely. He was fascinated by the Rembrandt painting of young David playing the harp while Saul—with fear and hatred in his face—looks on. There was a print of this painting in his office, and he had often given prints to colleagues to remind them that another generation was coming along. Emotionally, however, he had intensely dreaded retirement; and the actual event brought on a depression, the extent of which was apparent only to the immediate members of his family.

"I think he had really felt that he could go on forever," his son-in-law Jerome Berkowitz told me. "He was stunned by the fact that everybody at the institute simply accepted the idea that he would retire at seventy. I think he felt a little bit betrayed that a delegation hadn't come to him and demanded that he continue as director."

It was impossible for Courant to give up trying to guide what he had created. In addition, he was afraid, he told me, that Stoker "[might spoil] many things by his passionate and aggressive ways, which he has sometimes."

To take over the responsibility of directing the institute, Stoker had to shut Courant out until he had established himself as the director. This was not easy.

Besides having different views on the organization and financing of their mutual enterprise, Stoker found many aspects of Courant's personality very repugnant. Stoker is, as some of his friends say, "straightforward to a fault." He said quite frankly to me:

"There were things about Courant which I found detestable. Really very bad—and I can see why he had enemies, who couldn't see him as I could, what he could do positively. He was such a calculator. That was the worst. And his utter subservience to people who were rich—even if they were utterly boring and he wouldn't have anything to do with them if they weren't rich.

"But still, on the other hand," he said, his voice changing to one of great affection and admiration, "his merits were really so great. When I say those horrible things, I feel ashamed in a way because, after all, he created this affair. Under the most terrible difficulties. And he was so helpful to people of all kinds, me too—so willing to put himself out for people who weren't rich, because he had some feeling that they were worth the bother. And there were many of them.

"I don't think there's a better place anywhere in the world than this place. There is no such diversified group which is so cohesive. It still remains that way. The people are very cooperative. There are no internal bickerings or jealousies in the place. Just none at all. It's absolutely amazing. The worst trouble we ever had was between Courant and me. We did really clash, because, I think, I was the only one who was willing to stand up to him. In retrospect—well, that I should say—I regard Courant as a very great man. An absolutely tremendous man in his accomplishments."

By the time that Stoker became director, there was at NYU a core of outstanding mathematicians, among them nine future members of the National Academy of Sciences, who have remained until the present time in spite of tempting offers from other institutions. One of the harshest criticisms of Courant—voiced to me by Saunders MacLane of the University of Chicago—was that he hung on to the people he had trained and in that way actually hindered the development of applied mathematics in the country as a whole.

Stoker snorts at this criticism.

"MacLane and the others would have taken everybody Courant had trained, and where would we have gotten replacements? Nobody else was training them!"

Nevertheless, by the time Stoker succeeded Courant as director, a decision had already been made that there should be a greater effort to bring in people from the outside. Among those who came during the late fifties and early sixties were Jacob Schwartz, Paul Garabedian, and Monroe Donsker.

After I heard MacLane's criticism, I asked some of the second-generation people at the Courant Institute why they had stayed in spite of many financially tempting offers from other institutions. Their answers can be summed up in the statement that they like the place.

"Here, of course, most of us of my generation have been very well treated by Courant, socially, scientifically, and so on," Joe Keller told me. "That's not to say that our salaries were exorbitant. In fact, it was quite the contrary. But Courant made it nice for us in whatever ways he could. Somehow or other that made up for the salaries. Then we have all been pleased that we've had hardly any of the strife that is prominent in many other mathematics departments. An accompaniment of that, which is not independent of it, is that there is lots of mutual work—collaboration— which is not common in other mathematics departments. It is possible to ask someone for his help and then use it and not have to make him a coauthor."

"There are very few places in the country where you have this spectrum of pure and applied, and people can still talk to each other," Harold Grad contributed. "I have developed a theory. I am of the group you mentioned of my generation probably the most applied of them all while Louis Nirenberg —he is in mathematics straight, he doesn't do any applied mathematics. But maybe because we were both students together and I know that he respects me and I respect him, we figure that if the other one is doing something and we don't really know what he's doing, it must still be all right. In other places—of course I have never been anyplace else except to visit—I find that there seems to be only antagonism or disrespect for someone who isn't doing what you are doing. It is frequently considered a mistake to have kept so many people who were trained here, but maybe that's the reason we can talk to each other."

Since I had heard that the Courant Institute is a place where newcomers—and sometimes those not so new—feel that they are shut out from an "in group," I asked Grad if people who had not had the common training he spoke of could also communicate.

"Yes, communicate—no question. There are very subtle differences that in a sense you have to be born here to feel. There *is* a hierarchy—it's like China or Japan in the old days, where everybody knew exactly where he stood and what was going on. Almost impossible for an outsider. But if you ignore these subtle differences, which do exist, then communication is rather easy—just because our training has been to be able to talk to other people."

During Stoker's years as director, there was a spectacular expansion in the institute's research and graduate education program in mathematics. In 1961 a grant of $2,750,000 was made by the Alfred P. Sloan Foundation —this to be matched by an equal amount from the Ford Foundation and the National Science Foundation. Of the total, $4,000,000 went toward the construction of a new building for the Institute of Mathematical Sciences on ground already owned by the university. The other $1,500,000 went to support additional pre- and post-doctoral fellowships and new programs in statistics and mathematical physics as well as to establish a special $300,000 fund for research.

Sloan was eager to honor Warren Weaver, who after his retirement from the Rockefeller Foundation had gone to the Sloan Foundation. It was natural that Sloan would select the group at NYU for his tribute to Weaver's efforts on behalf of applied mathematics in the United States. For a few years after the war there had been plans at several important locations for substantial developments in applied mathematics; but, as Weaver himself wrote later in his autobiography, "it is distressing to have to record that in general these brave starts were not sustained." Since the war there had been only one "truly significant development"—and that had been at New York University.

Stoker and the other mathematicians at NYU did not object to their new building's being named Warren Weaver Hall, as Sloan suggested; but they did feel that the building should then also carry Courant's name. It had always been "Courant's institute" anyway. Although Courant could have objected—as Friedrichs pointed out to me—he did not.

"But I think that in a sense he was vaguely embarrassed by the whole thing," said Jane Richtmyer, an administrative assistant who acted as liaison between the faculty and the architects in the planning of the new building. "He loved the idea of his castle, but he also saw that it might lead to a breaking up or a loss of some of the things he valued. Certainly he was ambivalent."

Jane Richtmyer, the wife of Robert Richtmyer, the scientific director of the computing center after 1954, was another of Courant's "flames."

"He did like the ladies," she laughed when I talked to her in Boulder, Colorado, where she and her husband now live. "I used to think that he was a very 'human' human being with sort of an excess of everything. I never thought he was especially one way or another. He was both. Everything was just a little bit more than with everybody else. But basically he was just a very balanced, reasonable man with sort of excessive ways."

"Let me tell you about the time Papa felt that Jane was abandoning him," Lucille Wolff said to me during our interview. "He and I went for a long walk out in New Rochelle, very slow, inch by inch—he was quite old by then—and he told me how she was so cool to him and so on. 'Well, Richard,' I said, 'you know that this is not the first time you have been disappointed in love.' 'Ja,' he said, 'but maybe it will be the last time.'

"He dreaded the thought, you know—he loved the state of being infatuated. It was such a painful thing. When he was really in love, he was such a pest—you couldn't move without him being there. He was really a remarkable man, Papa—to be so distinguished and still to be so glad to humble himself. It made him feel again that he was true. It's hard to say. He was ambitious and humble. The more he wanted to be great, the humbler he got. And that was very charming."

In the planning of the new building, Jane Richtmyer worked with a committee of faculty members, all of whom had definite ideas about the needs of mathematicians. Friedrichs, for instance, wanted some *Gemüt-lichkeit*—in the library "a talking room" but also "a silence room" with easy chairs and paintings on the wall. Outside the auditorium and lecture halls on the main floor, he thought that there should be a sort of "temporary office" for preparation and relaxation before a talk.

Jane Richtmyer saw her job as translating these ideas and others to the architects "and trying to get them to value such things and make them practical."

On November 20, 1962, ground was broken for the new building. Courant and Weaver took turns holding the shovel for the newspaper photographers. Two years later the fourteen-story building was completed. By 1965 the mathematicians were in their new quarters.

I first met Courant shortly after the move to Warren Weaver Hall. I had recently returned to the United States from Germany, where I had visited the old institute on Bunsenstrasse, and I remember I could not help thinking that he had come a long way from Göttingen.

Visitors from Europe who have seen pictures of the building often say with confidence (although without success) to their New York taxi drivers: "Take me to the Courant Institute." It is a modest but elegant skyscraper, constructed of bronze-colored brick and glass. Powerful vertical columns rise thirteen floors to an imposing brick facade. A girdle of large bay windows around the twelfth floor provides the necessary horizontal contrast. Inside, nine floors of offices and lecture halls are carefully arranged on Courant's principle that "everything should be all mixed up." As an example, the thirteenth floor, which contains the office Courant occupied during his lifetime, also contains a large office for another senior professor, two lecture halls, a series of small offices for visitors and graduate fellows, and a long handsome lounge with couches and easy chairs where faculty members and students can enjoy coffee or tea throughout most of the day. The "second generation" professors sometimes tell stories of what it was like in the old days "at the Judson" and "at the Bible House" and, looking at the faces of the young people listening, suddenly see themselves again as students listening to Courant tell stories of Göttingen.

At the time I first met Courant, although he was seventy-seven years old, he was still very active and had only recently given up skiing. In addition to being director emeritus of the Courant Institute and a member of its board of governors, he served as scientific consultant for a number of large firms, most importantly IBM. Two years before, he had headed a delegation from the National Academy to the USSR's "academic city" near Novosibirsk. While I set up my tape recorder, he glanced constantly at his watch, rather like the White Rabbit, and murmured that he had to be in Washington that afternoon. I began our interview concerning his recollections of Hilbert by complimenting him on the handsome new building that bore his name.

"Ja," he mumbled, shuffling the pile of papers on his desk and glancing away. "Ja, ja, it is nice," he agreed. "But it does not guarantee achievement."

That same year Courant was asked by an interviewer what in mathematics he saw as most likely to be productive for the scientist.

"I don't want to emphasize or advocate future mathematical activities because of my personal taste," he said. "It is the same as in music. Even now I have difficulty appreciating Bartok or more modern music to any extent. Yet my grandchildren sit at the piano and play such pieces as a matter of course. They don't know there is any difference, and so will it be with respect to the attitude of the younger generation toward scientific subjects, such as computing and computers, or outlying fields of topology or logic. It may well be that older people are just no longer able to adequately absorb new material so I don't want to prophesy anything. I think as long as the attitude with which science is pursued is honest and not dominated by commercialism and as long as people are honestly dedicated, then one must have confidence that something valuable will result."

He still maintained a lively interest in writing. The year I met him (1965), he and Fritz John published in collaboration an "Americanization" of the first volume of the Courant calculus.

He himself had always intended to write a third volume of Courant-Hilbert after he retired. One of the subjects to be treated in it would be the application to practical numerical problems of the theoretical methods which had been developed in the earlier volumes. The other subject would be the existence of the solutions of elliptic partial differential equations. Already, in the English edition of the second volume (1962), he had omitted the seventh chapter on this subject, over which he and Friedrichs had struggled so long and valiantly in 1937. Now, in the third volume, he planned to rewrite the chapter and treat the subject afresh.

Over the years several young men were brought as visitors to work on the project. Among the first of these was De Prima. He found himself sadly unable to help Courant.

"The material for the third volume came from the old Courant-Hilbert, but it had changed and developed over the years," he explained to me. "Courant had really not kept up with the new results except in a general way. He needed to be educated so that he could understand the developments within his own framework. That occasioned some difficulties, in that Courant may have got beyond the point where he *could* understand. The questions he was asking were frequently naive and repetitive. Of course, we all know that in mathematics the naive questions are sometimes the most penetrating; and maybe we, the people trying to help him, myself included, did not quite sufficiently appreciate that.

"You see, whether it's abstract mathematics or concrete mathematics, you often push ahead by successive abstraction and generalization. Possibly what Courant was saying is that significant progress occurs only when you return again to ground level with new understanding. In earlier times it seems that Courant had a rare talent for posing simple and natural questions which often led to fruitful penetration of the how and the why of apparently abstruse and technically complicated mathematical situations. But now, it seemed, Courant was not really asking *those* questions. Maybe that was the trouble—perhaps he didn't know how to formulate them in the present context anymore—or perhaps he was too distracted in those last years."

In 1966, after eight years as director, Stoker resigned. During that time, with the assistance of Bromberg, his friend and former student, he had managed to achieve at last the long-desired recognition of the mathematics group as an integral part of the university. After Stoker resigned, Friedrichs accepted the directorship for a year. It was arranged that in the future the position would be assumed by different members of the staff on a rotating basis. Friedrichs's successor under this plan was Jürgen Moser, a pure mathematician whose work has nevertheless had a great deal of influence on the understanding of the peculiarities of the motion of asteroids.

Moser, who is the husband of Courant's daughter Gertrude, was born in Königsberg, the son of a neurologist. He attended the gymnasium at which Hilbert received the *Abitur*, and studied with Rellich and later with Siegel in Göttingen. Courant's other son-in-law, Jerome Berkowitz, Lori's husband, is also a mathematician and the chairman of the Graduate Division at the Courant Institute. He is typical of the students attending NYU when Courant came there. He is Jewish, a native New Yorker, a graduate of CCNY, the only son of a widow who had to work hard to support him after his father's early death. He came from a family who, as he says, "did not have quite the imagination for a career in mathematics, or in the university at all." Bromberg, for whom he was working at the Reeves Instrument Company after graduation, encouraged him to go and see Courant about continuing his education. Courant offered him a fellowship. Berkowitz's mother was a little upset at his giving up a job that paid $300 a month for one that paid $100, but Berkowitz knew somehow that it was all right. Today he says, "I think that very much of my education occurred right here, and I think that many of the people feel that way. The education was not just mathematical. It was political and social and musical, too."

The balancing and contrasting quality of Courant's sons-in-law is also exemplified by their wives, and by his sons and their wives. Courant's daughters carried on their careers with marriage. Gertrude became a biologist; and Lori, a professional musician. Not unexpectedly, Courant's sons became physicists rather than mathematicians. Ernest married Sara Paul, a young woman whose grandfather had been a rabbi. Hans married Maggie Spaulding, who had what Courant would describe as "a 100 per cent American background."

I came in contact with Courant again at the end of 1968 when he was reading the finished manuscript of *Hilbert*. He was eighty then, but he was trying hard to understand the alienation of the young whites, the anger of the young blacks. Because I came from the west coast, he questioned me closely about the Black Panthers. When I gave him Eldridge Cleaver's *Soul on Ice*, he would not let it out of his hands until he had finished reading it. But he was dreadfully unprepared and shaken in May 1970, during the Cambodian crisis, when a group of students took over Warren Weaver Hall and—announcing that they were holding the AEC computer for $100,000 ransom—covered the walls with obscenities and demands "to get the government out of the university."

All during the time that Moser was director, Courant was again tempted to meddle a little in the affairs of the institute. He made a confidant of young Monroe Donsker, a relative newcomer to the faculty. At lunches in the Village—insisting always on sitting against the wall "to avoid stabs in the back"—he expressed to Donsker his fears about the future of the institute and the weaknesses of his possible successors among the young people. He was most concerned that the institute was slipping away from its mission. In 1968—when he was eighty—he submitted to his colleagues "some remarks about the substance and continuity of our enterprise."

The role attributed to the institute as a center of applied mathematics must not be repudiated but cheerfully accepted. "It is, realistically seen, the brightest hope and justification for us in expecting . . . support from society, as our historical development clearly indicates, and it is a challenge that we must meet." There was a clear warning. "Before our eyes the process of more and more mathematical elements encroaching on more and more fields of scientific, technological, and other endeavors has been developing without finding in our institute the necessary degree of interested attention . . . the danger of losing contact with live developments has become real indeed."

During Courant's last years, the institute was in fact sliding somewhat toward pure mathematics. Louis Nirenberg, the "purest" of the second generation, had succeeded Moser as director. There was a feeling that the offerings should be broadened, more fields should be represented. The best students seemed to be attracted to the pure aspects of mathematics.

The struggle to write the third volume of Courant–Hilbert continued. The last young men drawn into it were Stefan Hildebrandt, a German visitor, and Don Ludwig, one of Courant's last Ph.D.'s. At one point Peter Lax suggested that Courant do the book in two small volumes, one with Hildebrandt and the other with Ludwig.

"I told Courant he wouldn't get it done any other way, and indeed he didn't," Lax said to me. "I don't know why he wouldn't do it. Maybe he thought he would be exploiting these people. He wouldn't have. I think it would have been a good opportunity for them, and something good would have come out of it."

Curiously, even in Courant's old age, when he and Hildebrandt began to work together, he conveyed to his young collaborator the same optimism in approaching mathematical problems which he had conveyed long ago in Göttingen.

"Technical difficulties—even if he was quite ignorant—never frightened him," Hildebrandt recalled. "He made one feel that any mathematical task is solvable provided that it is based on a sound and convincing idea.

After having spoken with him about a problem, I found myself always very confident that—in the end—the solution would be within reach."

In May 1971, in one of the last extended conversations I had with Courant, we returned again to the subject of the institute he had built up at NYU. It had been, he told me—psychologically as well as professionally—"a natural and intense" development for him:

"I felt so dedicated to doing something here for the benefit of the country and the best I could imagine that I might be able to do was to transfer my Göttingen experience and the skills and abilities I had acquired there to the situation here—this is what the institute downtown originated from. I also wanted to do something about the unity of science, including the mathematical sciences in this country. Of course, I consider that what we have done at our institute, we have achieved something; but we have not really won over people, have not achieved the continuation of some of the Hilbert tradition and the general tradition in Germany that was so close to my heart."

All during 1971 he was deeply depressed. The younger people at the institute continued to keep him informed about developments and to consider his wishes, but those who remembered the tremendous optimism and energy which had characterized him in the old days dreaded visiting him.

He no longer played the piano, and sometimes it seemed he could hardly bear the happy activity of others. Nina's passionate devotion to music continued. Her quartet evenings were regular events. Courant withdrew morosely to his study upstairs.

He became inordinately concerned with his financial situation. Yet even during his last years, in addition to his savings and investments, he had a more than generous yearly income which included consulting fees, royalties from his many books, a pension from the German government as well as the one from NYU.

"Look, Courant—I know—you *must* be a millionaire," Stoker remembers saying to him one time when he was fretting again about money.

"But it is so precarious, Jim," Courant said, "it is so precarious."

At NYU and other places, in 1971, the days of the Sputnik-inspired emphasis on science and the days of affluence were coming to an end. The last of the post-war babies who had thronged the colleges in the 1960's were being graduated. New York University, even more than most institutions, was feeling a financial squeeze which would shortly result in the forced sale of the entire University Heights campus to the City University of New York. As a result of the demand for "relevance," the claim of mathematics to be a fundamental academic discipline was being questioned in a way which was a throwback to the situation in America in the 1920's and 1930's.

The 83-year-old Courant worried incessantly about the future of the institute and of mathematics:

"I should go to Washington once more," he said wearily. "The new director of research for the National Science Foundation is a close friend of mine. But I cannot make up my mind to go and talk to him. You see, to do what I think should be done, everybody must be enthusiastic in the same direction. But gradually things have changed. It has become more and more difficult to deeply convince people of the need to do certain things. The situation of the world has changed too, and what should be done—in mathematics anyway—is no longer as clear as it was when I started. And so it is very difficult. Also there is a lot of unfriendliness and coolness toward us. I have had to make a long fight."

"You mean—here—in the beginning?"

"Always."

The last time I saw Courant was at the Courant Institute—the day before he suffered the stroke which resulted in his death in 1972.

A little more than three years later, I visited the Courant Institute again. Fritz John, who had been the last "non-Aryan" mathematician to obtain his Ph.D. in Göttingen in 1933, occupied Courant's old office. Peter Lax had succeeded Louis Nirenberg as director. In nominating Lax as a member of the National Academy in 1962, Courant had described him as "[embodying] as few others do . . . the unity of abstract mathematical analysis with the most concrete power in solving individual problems and with a well-balanced scientific philosophy." At the institute there was now the hope, often expressed to me, that perhaps—after a succession of directors who did not want the position but had assumed it out of a sense of duty—*Peter is the one.*

Lax feels that mathematics is flourishing today at the Courant Institute. True, there are the difficulties common to all universities and in all university mathematics departments—less money, fewer students—and it is no secret that New York University's financial situation is particularly desperate. There are also problems unique to the institute itself. The distinguished faculty members—the young people trained during the war years—are middle-aged now; and no "third generation" has appeared. On the other hand, the current economic situation—the disillusionment with knowledge that has no value in the job market—has worked to the advantage of a mathematics group as applied as the one at NYU.

Lax also feels that the institute is carrying on its mission—"to bring mathematics to applications—new applications—and to relate mathematics to science."

"This does not mean that we are continuing in exactly the same way," he told me. "We have modified to a certain extent, and we are continuing to modify. The institute today is more 'applied' than it has been for a long time."

He picked up a copy of the catalogue for the year 1974–75. On the cover there was a schlieren photograph of a shockless airfoil designed by Paul Garabedian, the principle of which has since been applied by one of his own students to the design of compressor blades and turbines for jet engines. The catalogue expresses, virtually unchanged, the philosophy which Courant espoused for so many years in so many memorandums.

"I am going to take this home and rewrite it a little," Lax said, fanning the pages under his thumb.

As I left Lax's office that day and entered a crowded elevator in which the other passengers—faculty members as well as students—were continuing a discussion begun upstairs in the lounge, I remembered the last time I had seen Courant—the last time he was in the institute that bears his name, the day before his stroke.

I had long since realized that the time for working with him on his reminiscences had passed, and I had given up trying to extract from him specific memories of people he had known and events he had experienced. That day I had been going over old letters and papers in his files. I had not realized that he was in the building, and I almost did not see him as I was leaving; for he was standing near his office, drawn back into a slight alcove so that students rushing past, most of whom no longer knew who he was, would not upset his precarious balance. Wearing a big fur-collared coat and a curly cossack's hat which he had bought in Moscow, he stood quietly, a small old man with a face like a miserable troll.

I stopped to greet him, shocked at how much he had failed since I had seen him last.

"I am a little tired and do not feel very enterprising," he murmured in an attempt at apology for his lack of response to what I told him about my work that day.

Seeking for something to say that would be cheering, I commented again how struck I was every time I entered the institute by the spirit, human and scientific, that I always felt there.

His eyes, which had been dull, brightened a little.

"Ja, ja," he nodded. "It is Göttingen. Göttingen is here."

INDEX

Göttingen
and
New York

AN ALBUM

I would like to express my appreciation to Natascha Brunswick and Nina Courant, who both furnished a number of pictures, and to each of the following for individual pictures: the Archiv für Kunst und Geschicht, Prof. Garrett Birkhoff, Elizabeth Schoenberg Brody, Nellie Friedrichs, Prof. Konrad Jacobs, Prof. Fritz John, Michael Lewy, *Life* Magazine (for the copyrighted photograph of Warren Weaver), Elisabeth Franck Lisco, the M.I.T. Historical Collections (for the picture of Norbert Weiner and Max Born), Prof. Jürgen Moser, Prof. George Pólya, Margaret Pryce-Born, Brigitte Rellich, Rockefeller University, Prof. Peter Swinnerton-Dyer and the Trinity College Library, Caroline Underwood and the School of Mathematics at the Institute for Advanced Study, Warren Weaver, F. Joachim Weyl, and Eva Toeplitz Wohl.

Top: Felix Klein / *Bottom:* Alfred Haar, Franz Hilbert, Minkowski, . . . , the Hilberts, Ernst Hellinger

Top: Courant as a student / *Bottom left:* Hermann Weyl / *Bottom right:* Nina Runge, later Courant

Top: Edmund Landau and daughter Dolli / *Bottom:* Otto Toeplitz

Top: G. H. Hardy / *Bottom left:* Carl T. Runge / *Bottom right:* Harald Bohr

Courant in the trenches, 1915

Top: James Franck (with Gustav Born) and Max Born / *Bottom left:* K. O. Friedrichs / *Bottom right:* Hans Lewy

Top: Courant in his first car, 1929 / *Bottom:* Otto Neugebauer

Top: David Hilbert from painted portrait / *Bottom left:* Weyl / *Bottom right:* Gustav Herglotz

Emmy Noether

Top: Norbert Wiener with Born / *Bottom left:* George D. Birkhoff / *Bottom right:* Oswald Veblen

Top: Courant, Landau, and Weyl / *Bottom left:* Carl Ludwig Siegel / *Bottom right:* Emil Artin

Abraham Flexner

Top: Courant and J. J. Stoker / *Bottom left:* Donald Flanders / *Bottom right:* Franz Rellich

Top: Franck and Born back in Göttingen / *Bottom left:* Fritz John / *Bottom right:* Warren Weaver

Top: Groundbreaking for Warren Weaver Hall / *Bottom:* Friedrichs and Courant

Courant in front of the Courant Institute of Mathematical Sciences, 1965